HELL CREEK, MONTANA

MONTANA

AMERICA'S KEY TO THE
PREHISTORIC PAST

LOWELL DINGUS

ST. MARTI... ...RESS

NEW YORK

www.stmartins.com

Library of Congress Cataloging-in-Publication Data

Dingus, Lowell.

  Hell Creek, Montana : America's key to the prehistoric past / Lowell Dingus.— 1st U.S. ed.

    p.   cm.

  Includes bibliographical references (page 233).

  ISBN 0-312-31393-4

  EAN 978-0312-31393-7

  1. Paleo-Indians—Montana—Missouri Breaks.   2. Indians of North America—Montana—Missouri Breaks—Antiquities.   3. Excavations (Archaeology)—Hell Creek Formation.   4. Paleontology—Hell Creek Formation.   5. Dinosaurs—Hell Creek Formation.   6. Frontier and pioneer life—Montana—Missouri Breaks.   7. Missouri Breaks (Mont.)—History.   8. Missouri Breaks (Mont.)—Antiquities.   9. Hell Creek Formation—Antiquities.   I. Title.

E78.M9D56   2004
978.6'2—dc22   2004040960

First Edition: July 2004

10   9   8   7   6   5   4   3   2   1

CONTENTS

ACKNOWLEDGMENTS

This narrative is not exclusively about me, although I play a supporting role. It's the story of a remarkable place and its hearty inhabitants. For the most part, I am simply the raconteur.

As such, I have relied on the research and skills of numerous others, including all those authors cited in the text and endnotes. But above all I'm indebted to the people of Jordan, Montana, especially those families and individuals who have nurtured and slaked my curiosity about their surroundings, including the Engdahls, Twitchells, FitzGeralds, and Herbolds. Beyond that, Jack MacRae, the president of the Garfield County Historical Society, provided invaluable assistance in both developing and researching many of the stories, as well as hosting a fabulous tour of his ranch and adjacent areas. Similarly, I owe a great debt to William Clemens, Harley Garbani, and all my other professional colleagues from UC Berkeley, who taught me about the scientific wonders of the region. Special thanks are due to Mike Greenwald,

Mark Goodwin, and Luis Chiappe, who contributed photos for this book. I also owe a great debt of gratitude to Jim Peters of the Parmly Billings Library, who made records of back-issue newspapers available for research and copying, following an introduction through John Russel.

The book itself would not have been possible without the steadfast encouragement and constructive criticism of my editors, Brad Wood and Ethan Friedman, along with my literary representative, Samuel Fleishman. Additional thorough and illuminating reviews by Elizabeth Chapman, William Madison, William Clemens, Harley Garbani, David Whistler, Larry Dingus, Stow Chapman, Judy Chapman, Eugene Gaffney, Gina Gould, and Willard Whitson greatly enhanced both the quality and fluency of the manuscript. Beyond that, Michael Novacek, Mark Norell, and Niles Eldredge were kind enough not only to read the manuscript but also to provide endorsements, for which I'm especially grateful. Karin Fittante aided greatly by persistently and creatively researching photos, and Jerome Greene most graciously allowed me to adapt his map of major battles in the Great Sioux War.

To all these contributors, I express my heartfelt gratitude and emphasize that any shortcomings, either in fact or style, are purely of my own making.

HELL CREEK, MONTANA

PROLOGUE

If you deem it worthwhile to delve deeply, down to the irreducible roots of this saga, you'll find that it's not really about Indians and soldiers, or pioneers and ranchers, or even dinosaurs and paleontologists, although I am one of the last. For all of these entities serve simply as thespians in the script that tells the true tale.

At times the world in which each of us acts seems static and stagnant. Just think of the times that you've said to yourself, "Life is so boring; nothing ever seems to change." Yet occasionally we are overwhelmed by a sudden event—a birth, a death in the family, a hurricane or earthquake, a war or a peace treaty. Even if we aren't directly entangled, such events remind us that life is not static, nor is the environment in which we exist. No matter the rate of change, whether it be gradually slow or catastrophically swift, if we are willing to entertain a more perceptive perspective, the truth becomes patently obvious. Change is all around us. It accrues constantly, affecting not only our personal lives, but more expansively, the very land on which we live. The whole world and everything in it evolves whether we like it or not. It's been so for billions of years.

It might seem that most radical change preferentially radiates out of cities. I live in New York. As host to the New York Stock Exchange and the United Nations, New York's restless residents refer to our home as the "Capital of the World." Hardly a day passes when some decision or event is not beamed round the globe, sending shock waves reverberating upon distant shores. The terrorist attack on the World Trade Center has drastically altered our global society, yet to think that New York monopolizes change on this Earth is myopic.

Even in Earth's most remote refugia, where time seems to assume serene states of arrest, change is actually constant. Take an apparently barren stretch of prairie and badlands in the outback of Montana, referred to by some as the Big Open or the Missouri Breaks. Through archaic eons and more modern eras, it's reveled in splendid obscurity.

Few regions harbor more pristine panoramas of the West than the Missouri Breaks around Hell Creek, near the small town of Jordan in east-central Montana. The terrain near Hell Creek encompasses rugged ravines sculpted forcefully into the Great Plains by the omnific Missouri River and its dendritic maze of tributaries. Along the jutting ridges formed by ancient sediment laid down in long-lost streams and seas, this land-

Looking northeast at the Hell Creek badlands on the Trumbo Ranch. The higher beds containing the layers of coal belong to the Tullock Formation, which were deposited after the large dinosaurs became extinct 65 million years ago. The lower beds at the base of the ravines belong to the Hell Creek Formation, which contains dinosaur fossils, including those of *Tyrannosaurus*. (LOWELL DINGUS)

scape documents the life-and-death struggles that its domains have hosted for unimaginable epochs.

More recent inhabitants often proclaim Jordan to be "the most isolated frontier town in the United States." There are only two paved streets serving this rural population, which numbers far less than a thousand. Nothing of global note must ever happen there, right? No! Not even Hell Creek can evade evolution. Quite cursory research reveals momentous events that are literally etched in the landscape.

Indeed, despite the remoteness of this territory and its tiny population, few locales have hosted more pivotal scenes in history, not only in terms of American history but in the history of life as a whole. Some events here have assumed almost mythic proportions within our social psyche, and many exploits of the actors are reminiscent of the adventures embodied by heroes and their adversaries in Greek and Roman mythology. This should not be too surprising, given the area's name, Hell Creek, but the people and sagas described in this book are anything but figments of an overactive imagination. They are figures fully ensconced in the very real pageant of evolutionary and human history. Moreover, the setting in which these characters played out these scenarios is truly epic in scope, with its treacherous badlands and seemingly endless prairies.

It would be difficult to overestimate the seminal nature of these events and participants. Bookshelves about them are heavily laden, but this tale will employ a more limited lens, focused primarily on the incidents at Hell Creek. Although these events are compellingly epic, amusing anecdotes often punctuate dire scenarios. So, to the maximum extent possible, the tales will be told in the actors' own phrases to preserve an atmosphere of authenticity.

Inevitably, my own sense of events will emerge, but that is not truly my point. I am a geological paleontologist, but most other pursuits lie beyond me: history, sociology, mythology, and agriculture to name a few in which I've indulged here. Feel free to oppose when it suits you; for, my goal is not to pontificate. There's plenty of that in the world.

My purpose is to portray the contrasting perspectives inher-

ent in these evolutionary dramas. For me, that's the essence of life, and the epic of Hell Creek simply mirrors the mysteries underlying most other experiences. Although the events have a true cause and effect, our ability to identify them is limited, and although the characters have true motivations, our ability to interpret them is, likewise, quite limited. In part this is due to the haze that clouds history. The farther one peers through the past, the less evidence one has to evaluate. In part, human nature intrudes, making stated motivations more suspect. Suffice it to say that Hell Creek has nurtured a perplexing array of life's ambiguities. Many characters and events are intensely complex and disdain resolution through simplistic analyses, a fact that has thoroughly entranced my curiosity.

Yet, there's one thing that can't be disputed. Seeds of historic upheaval have long taken root in these seemingly barren badlands and seasonally parched prairies. These convulsions have transformed our planet and all its biota, on scales ranging from continental configurations to singular souls. So, those forces and their effects have impacted much more than my life, making this tale much more than a personal memoir. Yet, despite this long saga, all that is clear is that more change is certain. Hell Creek and its inhabitants will continue to evolve.

Nonetheless, like the Native Americans and rugged ranchers of the region, my own life has not been immune to the forces of evolution at work in Hell Creek. The sprigs of those forces have entwined my existence through both the landscape and its people.

I conducted my research on the imposing scales of geologic and biologic evolution as a doctoral student among the buttes and coulees of the ranches at Hell Creek. I grew up in Los Angeles, and these summer field seasons afforded my first extended stays in the regions of wilderness where most fossils are found. So, Hell Creek served as my paleontological nursery. It was here that I first gained some fluency in the language that rocks really speak. Call me loony if you like, but I have come to value my conversations with rocks as highly as any gift that Nature has bestowed on me. The tales they have told me served as my license to explore the globe as a geological paleontologist on later expeditions to more exotic locales in Mongolia, China, and Argentina.

But beyond the expansive scientific tutorials that Hell Creek taught me about geologic time, the people there were also nurturing, in the near term of my day-to-day experience. The families of ranchers with whom we lived schooled me assiduously in the ways of their life and their land, all while personifying a toughness and resiliency that I came to deeply respect. Although I don't live at Hell Creek, I empathize profoundly with their plight. Because of my esteem for their sense of independence, I chose to base the narrative involving contemporary events primarily on the townspeople's public statements and the opinions that my friends freely expressed in the natural course of our conversations.

On the scale of those human souls, it is becoming increasingly difficult for the ranchers who nurtured me to make financial ends

meet in this periodically Hadean habitat, where summer temperatures rise over 110 degrees but winter brings chills below minus forty. Despite the proximity of the ample Missouri River, irrigation is, for the most part, nonexistent. So, many traditional family-run ranches in the northern Great Plains are failing, and the population of the region is decreasing drastically.

As a paleontologist, I'm acutely aware of the problems incurred when one peers through prehistory to interpret antiquity. So, predicting the future is way out of bounds. Yet, I can't resist wondering: Will Hell Creek's future be sculpted more by America's visions of approaching technologies or by the wild and remote vistas of its rapidly receding past?

1

BADLANDS AND BRANDING

I've been on the road from Berkeley for three long days and, as I head out of Jordan on the last stretch of pavement, I know I've abandoned my concept of civilization. While my rickety pickup cuts a steady course through gently rolling fields of desiccating wheat, the metropolis of Los Angeles, where I was raised, and the Bay area, where I go to school, lie far over the horizon toward the sunset. My attention is riveted to the road and the hand-rendered map my professor supplied. Four miles out of town the road jogs to the right, just past the Murnions' ranch. Two miles to the north, the pavement transforms into gravel that ricochets off the undercarriage. Three miles farther on it jogs to the left, before reaching the critical junction, a road bearing north to the Engdahl ranch, where I will spend the next six weeks during our summer field season. Small tracks lead off the wide swath of gravel into a checkerboard of pastures and grain fields, but which track should I take? I have no idea where I am in this labyrinthian landscape.[1] With evening descending, I wonder what Minotaur will materialize in this maze and if I can fend off its assault? About two miles past the last turn a sign looms up in the evening haze. Whitewashed boards are nailed to its vertical struts, so I stop and step out to investigate. They're all names: Baker, Hauso, Buffington, MacDonald, Lervick, on down to—there it is—Engdahl. Relieved, I turn north on a rut-pocked dirt track and head straight toward a date with the Fates.

Six miles on through the fields the road climbs a grade toward a small grayish knob on the skyline. Blinking back toward my map, I locate this landmark, modestly named Biscuit Butte. It's not very conspicuous by normal geographic standards, and I have no notion of what lies beyond. But, as I rumble on up toward the butte's silhouette, I can see that the road veers west. As I slow for the turn and rattle past a cattle guard a panorama of immense proportions materializes before me. I brake in stunned silence and take a deep breath. Dozens of square miles molded in pastel-colored badlands extend to the north, east, and west, as far as I can discern. Acutely sculpted ridges snake off toward the horizon alongside steeply incised ravines. Stalwart buttes, banded in rings of yellow, brown, and dark gray strata, rise resolutely above the chaotic incisions. Although grasses and shrubs cling desperately to the flat spots, few trees can withstand the harsh landscape.

Looking northwest at Biscuit Butte (on the horizon at left) and the Hell Creek badlands on the Engdahl Ranch. Most of the rock units seen below Biscuit Butte belong to the Tullock Formation. (LOWELL DINGUS)

Suddenly it dawns on me. This is Hell Creek, and the scene is indelibly etched in my mind. The vision is mesmerizing, and glimpses of long-lost worlds appear from below. A chorus of coyotes seems to chortle at me in the gathering dusk. I can almost imagine Cerberus sending his warning: Beasts beyond belief lurk in these badlands.

Nonetheless, I'm buoyed by an exhilarating sense of antici-pation. I'm a young, green paleontology student on my first expedition to this paleontological Mecca, in search of the first dinosaur fossils I will ever find. I stand on the precipice between prairie and badlands, straddling two entirely different worlds, a world of the present in which I live and a world of the past filled with remarkable monsters. Planted in the present, gazing down on the past as vague visions of *Tyrannosaurus* and *Triceratops* locked in mortal combat cavort through my mind. I sense that if I step off the plains' edge and descend to the badlands, I can transcend through time to the world of the dinosaurs.

After contemplating that world for a while I return to the present. Brief dusk is now shading the scene. I've still got to find the Engdahl ranch, and despite my musings of dinosaurs, my first actual encounter with the denizens of Hell Creek will have little to do with beasts of the past. In fact, that first tangible greeting will come from the modern fauna.

I've arrived a day earlier than the rest of the crew and need to check in with our hosts. They run cattle and sheep across thousands of acres of prairie and badlands. Lester and Cora are the patriarch and matriarch. At almost seventy, Lester still occasionally rides out on horseback to tend his sheep, although he usually now prefers the air-conditioned comforts of his four-wheel-drive "outfit." The icy winters and smouldering summers have tanned his crusty exterior, and although falls from frisky mounts have left him slightly stooped with a limp, he remains unbowed as he wanders the buttes of his ranch, always accompanied by his overly enthusiastic pup, Tippy. His son Bob and daughter-in-law Jane do most of the hard work around the ranch, assisted by their teenage children Cathy and Duane. Together they comprise a portrait that Norman Rockwell would have relished painting.

They invite me to dinner the evening I arrive. After a hearty meal of meat and potatoes, interspersed with occasional small talk, Bob turns his lanky yet powerful frame towards me and, with an intense look radiating from beneath the pale skin of his farmer-tanned forehead, pops a curious question.

"So, Lowell, you doin' anything about four-thirty tomorrow mornin'?" The question is accompanied by an ulterior look in his eyes.

Completely dumbfounded, I reply, "Not that I know of."

"Good," he twangs, "'cuz we got a few head to brand and we could use an extra hand. Shall I swing by the cabin and pick ya up?"

"OK," I respond sheepishly as I assess my five-nine and one-hundred-and-forty-pound frame. "But I'm not sure I can be of much use."

My anxiety is heightened when he explains with some urgency that, "We're a bit late for all this, and if we wait any longer those heifers'll be too big ta handle."

My night is almost entirely sleepless. The sky outside our camp cabin is resplendent with stars and planets and meteors. I watch them restlessly, wondering whether I can withstand the rigors of the next morning's ruckus. Although my dad was invited to try out for the Olympic decathlon and my older brother lettered in several sports, I am the runt of the family— the nerd. As a youngster, I loved to play pickup games with my mates, but I never played organized sports. I possessed neither the size nor the confidence. Yet, somehow I will have to measure up in the eyes of folks whom I've hardly met in an intimidating and unfamiliar arena. Finally, the glow of Bob's

The Engdahl family and the 1979 field crew from UC Berkeley. Top row, from left to right: Jane Engdahl, Robert Engdahl, Duane Engdahl, William Clemens, and Dave Archibald. Bottom row, from left to right: Mark Goodwin, Cathy Engdahl, Mike Greenwald, and Lowell Dingus. (PHOTO COURTESY OF MARK GOODWIN)

headlights appears over the adjacent hill. I feel the way a gladiator must have felt before he faced his moment of truth—with one significant distinction—at least he wouldn't die a humiliating death at the feet of a fledgling cow.

As we approach the corral my eyes grow as big as fully-plopped cow-pies. For calves, those suckers are truly huge, and they seem to be sizing me up. None weighs less than I, and some considerably more. I peevishly upbraid myself for not waiting to let my buddy, Mike Greenwald, a six-foot-two former strong safety, show up to take out some blockers. But Hell Creek does not grant such cowardly wishes.

Bob hops out, scrambles over the fence, and immediately begins to explain the routine. My job is to help Cathy and Duane wrestle and secure the calves while Bob brands them, vaccinates them, and castrates the ones that qualify. I'm petrified, but Cathy, a Scandinavian blonde who moonlights as a star guard on her high school basketball team and weighs less than I do, just gazes at me and giggles. She walks up and whispers her plan.

"Nothin' to it. I'll grab a front leg and help you hold 'em still, after you grab a back leg. Just hold on tight, but don't hurt 'em."

"Hurt *them*," I retort silently. "Christ, they must kick like jackhammers."

Unfazed, she saunters in front of a calf to occupy its attention, then motions for me to sneak up from behind. I get within range and reach down for a leg. Startled, the calf bucks and breaks free with a kick that grazes my wrist.

"Hold on with both hands," Cathy urges.

So, back to the melee I stagger, grimacing from the glancing blow. Once I get a firm grip, Cathy effortlessly grasps a front leg.

"Let's flip to your right," she suggests, but I can't conceive of just how. In desperation I swipe at the calf's other hind leg with my foot, and it tumbles down on its side with a thud.

"Whoa," Bob chuckles with glee. "Never seen that one before. Now, just stick yer foot up behind its knee and hold on while I get to work."

I tenaciously clutch the writhing hind leg as Bob blisters its

Branding at the Engdahl Ranch. Mike Greenwald, then a collections manager at UC Berkeley's Museum of Paleontology, pitches in by securing a calf while branding. (PHOTO COURTESY OF MIKE GREENWALD)

thigh with the brand. The calf's pupils roll back out of view as the brand singes its fur, and the acrid smoke wafts up my nostrils.

We brand 110 head before nine, and I feel eighty years old by the end. I'm plastered with dirt and brown slime. Every joint in my body aches, yet my sense of pride stands intact. I seem to have passed the audition. In some small way, I have won a part in the cast on Hell Creek's vast stage, but for the moment I perceive only one benefit: I have seen a cow's anatomy from every possible angle, so I will certainly ace that lab exam if the opportunity arises.

Over the next few days, as the field crew from Berkeley trickles in for the season, the paradox of the whole situation confounds me. Many of my colleagues sport the long hair and beards emblematic of our antiwar generation. Our appearance and views contrast starkly with those of most locals, but animosity between our crew and the ranchers is all but nonexistent, save for occasional needling.

I puzzle as to the source of this sincere respect between clearly contrasting constituencies. In part, it seems based on the value of mutual entertainment. Just as we are a welcome source of wonderment for the ranchers during their long, infernal summer chores, they and their lifestyle offer a much-needed sabbatical from the tedium and stress of the classroom. But beyond that, our mutual respect is forged through the challenges of surviving and succeeding in the harsh environs of Hell Creek, and although we "wrassle" dead dinosaurs rather than cattle, both pursuits are worthy of admiration.

Upon reflection, my first test in this hostile environment, through branding, seems like a fitting and welcoming initiation for joining the cast on the stage at Hell Creek. For the region has dished out far more trying tribulations to others in the past.

2

TIME TRAVELING THROUGH HELL CREEK

Today, most people see Hell Creek as one place, representing a slightly modified version of the Wild West, and after my antics while branding with the Engdahls, it's easy to see why this is so. But, to geologists and paleontologists, other views are revealed in the depths of the Breaks.

While the field crew assembles over the next few days, our professor, Bill Clemens, introduces us rookies to the area. He is tall and substantial in stature. But any potentially imposing effect is quickly cancelled out by his deeply pensive and resonant voice, and his full, bushy beard only enhances his air of calm wisdom. From my perspective, through Bill's experienced eyes, the bumpy tour through the Breaks in his creaky green Blazer is a chance to immerse myself in the time-traveling journey I imagined on the precipice of the badlands by Biscuit Butte.

The rock layers of Hell Creek comprise a chronicle of the

locale's geologic history. The stories contained in the rocky pages constitute epic poems about long-lost landscapes and their daunting denizens. By traversing the layers of rock, one embarks on an odyssey back through time, filled with geologic and evolutionary scenes that remain locked in the landscape.

The sequence of rock layers, deposited one on top of the other, is essential for telling time back when long-extinct organisms roamed Hell Creek's ancient environments. In a sense, each layer represents a page in the book of history for the region. By applying the geologic principle of superposition, we can leaf through epic scenes page by page—from younger, higher layers into older, lower layers. Thus, fossils found in the higher rock layers represent plants and animals that lived after fossils found in lower rock layers. So, each fossiliferous layer contains distinct events in the saga of life at Hell Creek, because each rock layer represents a different verse or moment in the geologic and evolutionary history of the area.

The scenic drive from Jordan about twenty-five miles to Hell Creek State Park provides our crew with a transect for geologic time travel. And through Bill's expert eyes, dazzling impressions of ancient domains begin to emerge in detail beyond my perception. It's as though Bill is a Delphic oracle, peering intensely into the misty past, rather than into the hazy future.

The Hell Creek badlands, looking north from the road to Hell Creek State Park. The beds near the top of the sequence belong to the dinosaur-bearing Hell Creek Formation. Near the bottom of the bluffs, beds of the Fox Hills Formation and the marine Bearpaw Shale are exposed. (LOWELL DINGUS)

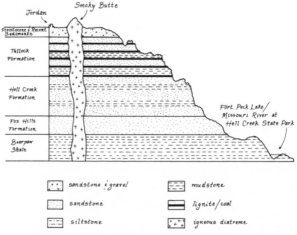

Schematic cross-section of the rock units exposed along the road from Jordan, Montana, to Hell Creek State Park. The rock formations are named on the left. A key for the basic rock types found in these formations is provided below the cross-section. (LOWELL DINGUS)

Present to 57 Million Years Ago

As we head north from Jordan, the first leg is laced with the same agricultural fields that I first drove through two days ago on my way to the Engdahls. The fields, with their fertile brown topsoil, contain sediment eroded from rocks deposited by streams soon after *Tyrannosaurus* went extinct, 65 million years ago. More sediment washed out of the glaciers that scoured the landscape just north of Hell Creek within the last few hundred thousand years during the ice ages.

With Bill as our farsighted guide, we see in the distance a prominent geologic feature peeking out of the prairie about eight miles east of Jordan. Long referred to as Smoky Butte because of its dark gray color, it rises to a conical peak several hundred feet above the surrounding prairie. Formed about 50 million years ago when a gas-laden body of molten magma forced its way up through the surrounding sediments from deep within the Earth's mantle, it baked the adjacent layers of strata and eventually cooled into a hard, dark gray rock called a diatreme.[1] Although no one has ever witnessed the formation of a diatreme, it is thought that as the body nears the surface, the gas expands as it does when one opens a champagne bottle, resulting in a violent volcanic eruption. Because of the hardness of the rock that forms the diatreme, it erodes more slowly than the softer sediments into which it intruded, which explains why it now forms a butte. In other areas of central Montana, diatremes are thought to produce a residue of sapphires, and possibly even diamonds, after other minerals in the diatreme eventually weather away into clay. But Smoky Butte contains an unusual mineral called armacolite, which is found nowhere else on Earth. In fact, the only other rocks known to contain this mineral were collected by Apollo astronauts on the moon.

57–65 Million Years Ago

As the road reaches the edge of the badlands that form the Missouri Breaks, we descend to the depths of Hell Creek, dropping off the edge of the modern world into earlier prehistoric landscapes and seascapes rooted deep in geologic time.

We pass through yellow and grayish layers of sand and silt laid down by rivers and streams on a gently sloping, swampy floodplain. This subtropical ecosystem was populated not with dinosaurs, but by a rich fauna, which included some of our own distant mammalian relatives that succeeded the dinosaurs. Today termed the Tullock Formation, these rocky strata entomb the earliest exploits of evolution during the Age of Mammals. Most identifiable fossils of these primitive mammals consist of isolated teeth, and most of the teeth are quite small. None of these mammals were larger than a beaver, and most were no larger than mice. It would be another 10 million years, after the extinction of large dinosaurs, before mammals would evolve into the larger body sizes that are commonly seen today.

In order to find these minute mammalian teeth, which are

Bill's paleontological specialty, one usually needs to get down on one's hands and knees to crawl slowly over the ground. The most common place to find such small fossils is near the bottom of sandy channel deposits that represent ancient dunes and sandbars that filled the streams that ran through the swamps. Since the fossils were slightly larger and heavier than most of the sand grains, the currents in the streams tended to tumble them along the stream bed where they accumulated in the low spots. Away from these stream channels, lush vegetation filled the swamps with decaying plant material, which, after being buried, baked, and compacted for more than 60 million years, formed the seams of dirty coal that crop out near the top of the badlands.

Dipping down into a small basin, we reach the lowest layer of coal in the sequence, the moment in geologic time when the last large dinosaurs disappeared. In deference to its basal position among all the other coal seams in the region, it is named the Z-coal and has attained a fame all its own. For no dinosaurs were preserved above, or after, this layer of coal formed. But their skeletons were entombed by the thousands in the few hundred feet of rock layers deposited below, or before, the level of this coal.

65–67 Million Years Ago

Descending past the Z-coal we encounter a wonderland of pastel-colored greenish, purplish, bluish, reddish, and brownish sandstones and siltstones. To me, the spectacularly sculpted outcrops resemble natural palaces of Olympian proportions. These touchstones of evolutionary innovation were named the Hell Creek Formation by the pioneering paleontologist, Barnum Brown. They house history from a time before 65 million years ago, when dinosaurs inhabited Hell Creek.

The environment then was strikingly different from that of today. As Bill explains, vast rivers and small streams meandered across a steamy, gently sloping floodplain that supported forests of conifers and lushly vegetated but fairly well-drained swamps. The ancestral Rocky Mountains lay about a hundred miles to the west, and a shallow inland sea bordered the floodplain to the east. This ancient subtropical paradise played host to algae,

Artist's depiction of the Hell Creek environment 65 million years ago, when *Triceratops* and other dinosaurs roamed the ancient floodplain. (PAINTING BY ELEANOR KISH; REPRODUCED WITH PERMISSION OF THE CANADIAN MUSEUM OF NATURE)

clams, snails, fish, frogs, turtles, lizards, snakes, crocodiles, birds, and more primitive, minuscule mammals, as well as dinosaurs. The kinds of fossil plants found here, coupled with the presence of animals like crocodiles, document that the environment was quite equable. The days weren't too hot, nor the nights too cold. The summers weren't too torrid, nor the winters too frigid. The climate of the Earth as a whole was more uniform, as evidenced by the lack of polar ice caps at that time, and the shallow seaway acted like a hot-water bottle to buffer the floodplain against climatic extremes.

Within this densely populated habitat, dinosaurs were the most conspicuous carnivores and herbivores. In addition to the fearsome *Tyrannosaurus* and its slightly smaller cousin, *Albertosaurus*, the portrait that Bill paints includes diminutive meateaters, such as *Dromaeosaurus* and *Troödon*, which scurried through the underbrush. These tiny predators, only a few feet long, darted nimbly about on their two hind legs in search of prey. Although distant relatives of *Tyrannosaurus*, they were actually much more closely related to the group of dinosaurs that still survives today—birds. On a slightly larger scale, the ostrich-like *Ornithomimus* reached a length of about fifteen feet, but was not as imposing as its larger dinosaurian cousins, due to the lack of teeth in its mouth.

In terms of herbivorous inhabitants, a couple kinds of armored dinosaurs wandered the Hell Creek floodplain, including the tanklike *Edmontonia* and *Ankylosaurus*. Horned dinosaurs that munched on the lush vegetation included the imposing *Triceratops* and *Torosaurus*, with their intimidating array of bony horns and studded shields on their skulls. Duckbills also inhabited Hell Creek, as shown by the fossil remains of two genera, *Anatotitan* and *Edmontosaurus*. Finally, Bill describes three types of dome-headed dinosaurs that filled out the domain: the imposing *Pachycephalosaurus,* along with its smaller cousins, *Stegoceras* and *Stygimoloch.*

67–68 Million Years Ago

As our descent continues below these dinosaur-bearing layers, the road meanders through weathered exposures of yellow-gray sand, which herald a major evolutionary shift in the ancient environment of Hell Creek. Although the summer sun now bakes these sediments, Bill's Chevy Blazer is actually four-wheeling across the shoreline sands of the shallow ancient seaway that extended off to the east of Hell Creek. Attesting to the shoreline environment that these sands represent are fossils of clams, snails, plants, and other brackish- to shallow-water organisms.

68–69 Million Years Ago

But, before long Bill leads us into a landscape dominated by low, rounded hills of brownish-gray, silty mudstone. We had entered the murky realm of the Bearpaw Shale, sediments deposited at the bottom of the shallow seaway that from time to time during the Age of Dinosaurs extended all the way from the region around today's Gulf of Mexico up to the polar waters of the Arctic Ocean. This seaway teemed with exotic menageries. Stupendous reptiles, such as the imposing marine lizards called mosasaurs and the long-necked, sea-serpent-like plesiosaurs, prowled the waters. Even relatives of our modern-day nautilus, called ammonites, flashed through the sea with their gleaming, pearly shells, and microscopic plankton with intricately twisting tests of calcite passively rode on the currents and formed the base of the seaway's food chain.

Before 70 Million Years Ago

As we reach Hell Creek State Park, the road descends to the lowest rock layers of Bearpaw Shale exposed in the area. Here the dammed waters of the modern Missouri lap up against the base of the badlands. But rocky cores drilled from oil wells that dot the region have revealed a thick sequence of layers buried beneath the surface. They record even more remote chapters in Hell Creek's long geological legacy.

The period preceding the reign of *T. rex* and the mosasaurs marked the rise of the Rocky Mountains. Their uplift, between 70 million and 90 million years ago, resulted from gargantuan collisions between the light continental plate in the Earth's crust that forms North America and the heavier oceanic plate that forms the bottom of the Pacific Ocean.[2] These prodigious plates were driven together by the immense forces of plate tectonics operating deep within the planet. As the westward moving North American plate plowed into the Pacific plate, the denser oceanic plate dove under the western edge of North America, which at the time, ran from northeastern Washington, through western Idaho and eastern Oregon, down into the Sierra Nevada region of California. As the Pacific plate descended under the western coast, tremendous temperatures and pressures melted the bottom of the North American continent, generating monumental masses of relatively light magma, which rose to form great aggregations of granite under the old coastline. As heat drove the rocks buried beneath the surface to expand, the western margin of the continent bulged upward, giving rise to the Rocky Mountains. It was the sediments eroded from these ancestral Rockies that formed the floodplain domain of the dinosaurs at Hell Creek.

Yet the plate tectonic motions that manufactured a distinct North American continent originated even earlier, about 200 million years ago, when a rift opened up in the supercontinent called Pangaea, which incorporated all of the Earth's present-day land masses. This rift ruptured as dense, molten magma from the Earth's mantle rose and diverged under Pangaea, splitting the plate and opening the Atlantic Ocean. To the west of this rift, North America began its westward migration, which initiated the collision between the old western coast and the Pacific plate. This compressed the rocks that formed the western coast, thickening the continental margin and creating a broad highland.

Sediments eroding from the eastern flank of this highland

formed rock layers, called the Morrison Formation, now buried beneath Hell Creek. These rocks attest to an age about 150 million years ago when massive dinosaurs dominated the continents. Supreme among these, in terms of size, were the sauropods, four-legged, long-necked behemoths. Commonly known genera include *Apatosaurus* (once called *Brontosaurus*) as well as *Diplodocus*, and *Camarasaurus*. But the largest of the Morrison's menagerie was appropriately named *Seismosaurus*. This beast literally shook the earth when it walked with its bulky body, which approached a hundred feet in length and fifty tons in weight. Sharing the Jurassic environment were both large and small carnivorous dinosaurs, including the imposing *Allosaurus* and the nimble *Ornitholestes*. Armored dinosaurs, such as the plated *Stegosaurus*, also roamed the riparion environment, vegetated with early relatives of conifers, cycads, ginkgos, and ferns.

Buried deeper beneath the surface lie layers of limestone called the Madison Formation, filled with fabulous fossils from 350 million years ago. These organisms also lived in an ancient sea that enveloped the landscape, but the tropical tides of this sea supported reef communities, dense with dazzling corals, delicate crinoids (relatives of modern sea lilies), and brachiopods with intricately sculpted shells.

In parts of western Montana geologists and drillers discovered other rock layers below this limestone formed from even more ancient sand and mud. They are called the Belt Formations because geologists first recognized and studied them where they crop out on the surface in the Belt Mountains, just east of Helena. These layers of sand and mud eroded off a highland to the south between about 1.5 billion and 800 million years ago. Although they contain fossils of microscopic algae and other single-celled organisms, no fossils of multicellular animals are present, because they do not appear in the fossil record until more than 200 million years later.

The rocks of the Belt Formation extend down to the very foundation of the North American continent, a mass of granite, gneiss and schist that formed about 2.7 billion years ago in the molten cauldrons of the early Earth. Geologists term this assemblage of igneous and metamorphic formations "basement rock," for, in a very real sense, it forms the basement of the continent on which we live.

In all, the North American continent is about twenty-five miles thick. So the thousand feet of floodplain and seaway sediments that we find exposed on the surface represents only a thin skin on the continental body and only the latest few scenes in the geologic history of Hell Creek and its inhabitants. For me, seeing this sequence of rocks for the first time, Bill's tour is at least as adventurous and challenging as any mythological odyssey.

But, long before my passage through these daunting breaks along the Missouri River, other more intrepid explorers braved a truly pristine wilderness in a journey that would forever alter the course of both American and human history.

3

EARLY EXPLORATION

By 1805 the environment around Hell Creek looked much as it does today. However, nomadic tribes of Native Americans roamed the Breaks, frequently confronting grizzlies and hunting immense herds of bison that migrated across the adjacent prairies.

It was across this formidable frontier that Thomas Jefferson sent the Corps of Discovery, led by Meriwether Lewis and William Clark, to assess newly acquired lands drained by the Missouri River. The names of these men still stand firmly atop the pantheon of American heroes, in no small part for the contributions they made to this pivotal project that opened up the American West to frontiersmen and pioneers. Less commonly known, however, is how the experiences of Lewis and Clark around Hell Creek endangered the expedition and almost cost them their lives.

Above all, Jefferson was a man of diverse interests and

knowledge. As the nation's third president, twenty-five years after authoring the Declaration of Independence, one of his principal accomplishments was the expansion of the fledgling nation's western border through the purchase of land from Napoleon that was referred to as Louisiana. At that time this broad swath of land essentially encompassed the territory drained by the Missouri River between the Mississippi River and the Rocky Mountains, as well as the southwestern tributaries of the Mississippi.

Even before the Louisiana Purchase was executed, Jefferson was intent on surveying it. His goals were both economic and strategic. As Stephen Ambrose notes in *Undaunted Courage*,[1] Britain, Russia, Spain, and France had all established a stake in the region. Britain ran a network for trading furs in what is now Canada, as well as claiming the Oregon territory. Russia operated commercial ventures near the mouth of the Columbia River and the region to the north. Spanish explorers and missionaries operated all along the Pacific Coast, while the French and French Canadians had opened up numerous areas along the Mississippi River and were contemplating the possibility of reestablishing their control over all the Louisiana Territory.

These claims of foreign dominance on the American continent directly conflicted with Jefferson's vision for his country. He wanted to establish an "Empire of Liberty" throughout, arguing that "Our confederacy must be viewed as the nest from which all America, North or South, is to be peopled. . . ."[2]

In addition to augmenting his own prestige and that of his

country, he recoiled at the thought of the continent being sliced up into European-style nation states. Having spent the last quarter century helping to establish a new democratic country, Jefferson naturally wished to extend the ideals of freedom embodied by the American Revolution from coast to coast.

But much of the Louisiana Territory was essentially unexplored. No reliable maps even existed for the waterways that ran through the region, and Jefferson needed someone to ascertain what was really there.

Based on their common roots grounded in the countryside of Virginia, Jefferson thought he knew the right man for the job of surveying the strategic, economic, demographic, and scientific assets of the Louisiana Territory—his compatriot, Captain Meriwether Lewis. Soon after his election Jefferson enlisted Lewis to be his secretary, with the ulterior motive in mind to prepare Lewis for the expedition. Lewis had an intimate knowledge of both the army—which would form the organizational foundation for the expedition—and of life on the frontier. These assets derived both from his military experience, when he traveled extensively west of the Appalachians, and his childhood in Virginia and Georgia, where he reveled in exploring the areas around his homes. Jefferson's tutorial took more than two years, from April 1801 to July 1803, during which Lewis learned how to describe scientifically the lands and biota that he would encounter. For a scientific inventory formed a vital part of Jefferson's vision for the mission.

That mission was comprehensively laid out by Jefferson in

his instructions to Lewis, finalized on June 20, 1803. The primary purpose was summarized in one paragraph:

> The object of your mission is to explore the Missouri river, & such principal stream of it, as, by it's course & communication with the waters of the Pacific Ocean, may offer the most direct & practicable water communication across this continent, for the purposes of commerce.[3]

So their glimmering goal was to find a freshwater conduit for commerce analogous to the Northwest Passage long sought by earlier European explorers. But mapping and gathering scientific information along the way was seen to comprise an essential and continual focus:

> Other object[s] worthy of notice will be the soil and face of the country, it's growth and vegetable productions; especially those not of the U.S.
>
> the animals of the country generally & especially those not known in the U.S.
>
> the remains and accounts of any which may deemed rare or extinct
>
> the mineral productions of every kind. . . . [4]

Jefferson's interest in scientific matters transcended the needs of developing the country's resources. He is known to have been an avid inventor, as well as an amateur paleontologist.[5] As a member and president of the American Philosophical Society, Jefferson helped raise funds to initiate paleontological expeditions and to investigate early fossil finds. One such enterprise involved a collection of fossilized bones discovered in Greenbriar County, Virginia. They were delivered to Jefferson's home at Monticello, where he conscientiously worked to describe and identify them. On his way to assume the vice presidency in 1797, he delivered the fossils to the Society, along with a report entitled, "A Memoir of the Discovery of Certain Bones of an Unknown Quadruped, of the Clawed Kind, in the Western Part of Virginia." Upon further study, they were identified to be the remains of the first giant ground sloth ever discovered in North America, a species that lived during the ice ages and to this day bears Jefferson's name—*Megalonyx jeffersoni*. Unbeknownst to either Jefferson or Lewis, the prescribed route along the Missouri River that the Corps of Discovery would follow would take them right through one of the richest dinosaur graveyards on Earth.

Lewis assembled a heroic crew of the frontier's best hunters and woodsmen, just as Jason assembled a legion of Greece's most legendary heroes. He chose for his counterpart on the expedition a military comrade with extensive knowledge and experience on the American frontier west of the Appalachians—William Clark. Together, they selected the other members of the expedition as they headed down the Ohio River, then up the Mississippi to St. Louis, between July and December of 1803. They were especially bouyed by the news reported on

July 4, 1803, that Jefferson's chief negotiator, James Monroe, had succeeded in his mission. Napoleon had agreed to sell all of the 800,000-square-mile Louisiana Territory to the United States for about $15 million—or roughly three cents an acre. From the Corps' point of view as American citizens, they would now be surveying their own sovereign land.

Lewis and Clark spent the winter and spring months of 1803–1804 preparing their plans and purchasing supplies while stationed near St. Louis at Camp Dubois, across the Mississippi from the mouth of the Missouri. Finally, in late May, they were on their way up the Missouri. By late October 1804 the crew had navigated on the river through the present-day state of Missouri, along the eastern borders of Kansas and Nebraska, through South Dakota and most of the way through North Dakota, where they wintered with the tribe of Mandans until April 7, 1805. A few key members of the crew were recruited at their hospitable haven among the Mandans, including a French Canadian trader named Toussaint Charbonneau, who lived with the Hidatsa tribe, and his Shoshone wife, Sacagawea.

With the preparations complete and the crew fully assembled, Lewis and Clark set out for an unknown fate on the frontier. By May 3 the expedition was two thousand miles upriver from the mouth of the Missouri, deep within uncharted territory and approaching the eastern edge of the daunting badlands now known as the Missouri Breaks including Hell Creek. The Corps would also be forced to face perilous predators every bit as belligerent as the fire-breathing bulls that the Argonauts engaged. Especially perilous were their encounters with grizzlies. Although the Native Americans of the region had warned the Corps about these immense bears before they ever encountered one, the crew arrogantly believed that their superior firearms would make quick work of these powerful predators. They even looked forward to their first run-in with them. Although they had encountered a few grizzlies farther down the Missouri, none of those confrontations had served to dampen the crew's reckless sense of superiority. But on May 5, in the region between today's Redwater and Milk Rivers, as Clark recounts:

> In the evening we saw a Brown or Grisley beare on a sand beech, I went out with one man Geo Drewyer & Killed the bear, which was verry large and a turrible looking animal, which we found verry hard to kill we Shot ten Balls into him before we killed him, &5 of those balls through his lights This animal is the largest of the carnivorous kind I ever saw . . ."[6]

Lewis elaborated that

> it was a most tremendous looking animal, and extremely hard to kill notwithstanding he had five balls through his lungs and five others in various parts he swam more than half the distance across the river to a sandbar, & it was at least twenty minutes before he died; he did not attempt to attack, but fled and made the most tremendous roar-

ing from the moment he was shot. We had no means of weighing this monster; Capt. Clark thought he would weigh 500 lbs. for my own part I think the estimate too small by 100 lbs. he measured 8. Feet 7½ Inches from the nose to the extremety of the hind feet. . . . [7]

After spotting another grizzly swimming across the river the next day, Lewis noted a marked change in attitude among the crew:

I find that the curiosity of our party is pretty well satisfied with rispect to this animal, the formidable appearance of the male bear killed on the 5th added to the difficulty with which they die when even shot through the vital parts has staggered the resolution of several of them, others however seem keen for action with the bear; I expect these gentle-men will give us some amusement shotly [shortly] as they (the bears) soon begin now to coppolate. [8]

How prophetic Lewis was. Between May 8 and May 20, 1805, the expedition struggled up the Missouri, passing the mouth of Hell Creek and traveling directly through the field area where we paleontologists work today. On May 8, as the crew passed the Milk River and entered the Hell Creek region proper, Lewis described two plants that were new to science, wild liquorice *(Glycyrrhiza lepidota)* and breadroot *(Psoralea esculenta)*. As Clark notes in his journal, Sacagawea gathered both

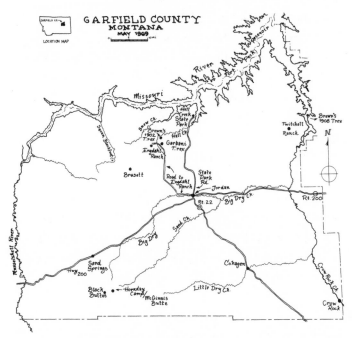

Garfield County, Montana, showing location of major sites related to incidents discussed in this narrative. (REDRAWN FROM "GARFIELD COUNTY, MAY 1969," INSIDE BACK COVER OF *TRAILIN' THROUGH TIME*, THE GARFIELD COUNTY HISTORICAL SOCIETY, 1999)

of these plants, so although the plants were new to science, they were well known to the Native Americans of the region and commonly used as a source of food. Although the scenery was inspiring as well as breathtaking, they had descended into a wilderness replete with intimidating denizens and conditions. Indeed, their journals describe some of the most momentous and perilous events that punctuated their two-year-long trial of endurance. In 1999 Jack McRae, president of the Garfield County Historical Society, compiled the accounts of the Corps of Discovery during their transit along the northern edge of the future county containing Hell Creek. This summary is included in the Historical Society's publication commemorating the centennial anniversary of the founding of Jordan, entitled *Trailin' Through Time*.[9]

On May 9 Lewis recounts that:

> today we passed the bed of the most extraordinary river that I ever beheld. It is as wide as the Missouri is at this place or one half mile wide and not containing a single drop of running water . . . This stream (if such it can properly be termed) we called Big dry River.[10]

It was a name that would stick to the present day, and many momentous events in the history of both Garfield County and the nation as a whole would eventually unfold along its banks. Lewis also notes that the area teemed with herds of bison and elk, and that the bison were "so gentle that the men frequently throw sticks and stones at them in order to drive them out of the way." By the end of the day they had traveled more than twenty-four miles and set camp at what is now called Duck Creek in Valley County. The Missouri now forms the dividing line between Garfield County to the south and Valley County to the north, so the fact that they camped in the latter simply means that they chose to camp on the Missouri's northern bank.

Lewis also killed and described four specimens of a new species of bird, now called the willet:

> It resembles the grey or whistling plover more than any other of this family of birds, tho' it is much larger ... the legs are of a greenish brown; the toes, three and one high at the heel unconnected with a web, the breast and belly of a brownish white; the head neck upper part of the body and coverts of the wings are of a dove coloured brown which when the bird is at rest is the predomanent colour. the tale has 12 feathers of the same length (nearly) of which the two in the center are black with transverse bars of yellowish brown, the others are a brownish white. the large feathers of the wings are white tiped [tipped] with black. the eyes are black with a small ring of dark yellowish brown—the beak is black, 2½ inches long, cilindrical, streight, and roundly or blountly pointed; the notes of this bird are louder and more various than of any other (kind of) species which I have seen."[11]

The amazing array of wildlife continued to catch the attention of the Corps on the following day. Their progress was hampered by high wind, so much of the day was spent hunting for camp food. Their quarry included two mule deer, one longtailed deer, two bison, and five beaver, but in addition, they encountered several elk and three bighorn sheep.

A willet *(Catoptrophorus semipalmatus)* shown flying through its habitat in Montana.
(DOUG WECHSLER, VIREO IMAGES)

May 11 brought the crew their first encounter with a grizzly in the region immediately adjacent to Hell Creek. After a cold and frosty morning, during which Clark discovered the distant hills to be forested with pitch pine and dwarf cedar, the late afternoon's activities were punctuated by Private William Bratton's frantic shouts as he sprinted toward his compatriots. As they met him at the shore, he related that he had shot a grizzly, which although too badly wounded to overtake him, nonetheless turned on him. Clark assembled a contingent with seven reinforcements, who all set out to track the beast down through a mile of dense rose bushes and willows. The grizzly was finally dispatched with two bullets through the skull, but in slaughtering it, they realized that Bratton's original shot had passed right through both lungs without killing it. After the bear survived for two hours with such a normally fatal wound, Lewis confided to his journal:

> These bear being so hard to die reather intimedates us all; I must confess that I do not like the gentlemen and had reather fight two Indians than one bear.[12]

Over the next two days, through intermittent rain and high wind, the Corps covered more than twenty-five miles, as Lewis scientifically described the flora, including the blooming chokecherries. At the end of the day on the thirteenth they camped just upstream from the mouth of what is today called Crooked Creek in Garfield County.

On May 14 the Corps traversed through the heart of the Hell Creek region. Passing the mouths of two dry streams along the southern bank, they named them Stick Lodge Creek and Brown Bear Defeated Creek. Today they are respectively known as Hell Creek and Snow Creek. The name attached to Brown Bear Defeated Creek was no figment of the explorers' imagination. Six of the men in the two trailing canoes, "all good hunters," discovered an enormous grizzly lying in open ground and decided to attack it. Creeping within forty yards of him, "unperceived," four of the men fired their rifles, which resulted in two of the bullets passing "through the bulk of both lobes of his lungs." Much to their amazement, not to mention terror, the grizzly, thoroughly infuriated at their attack, charged at full tilt. Despite receiving two more wounds from the other two riflemen, one

Painting depicting the confrontation between a grizzly and the crew of Lewis and Clark entitled "Hasty Retreat." The incident occurred along the Missouri River in the Hell Creek region on May 14, 1805. (PAINTING BY JOHN FORD CLYMER IN THE COLLECTION OF THE CLYMER MUSEUM OF ART, ELLENSBURG, WASHINGTON)

of which broke its shoulder, the enraged bear continued its charge, forcing the men to take flight for the river. By the time they reached the river, the grizzly had all but overtaken them. Two crewmen took to the canoe, while the others took refuge in the nearby willows and fired more balls into the bear,

> but the guns served only to direct the bear to them, . . . he pursued two of them separately so close that they were obliged to throw aside their guns . . . and throw themselves into the river although the bank was nearly twenty feet (high); so enraged was this animal that he

plunged into the river only a few feet behind the second man . . . when one of those who remained on the shore shot him through the head and finally killed him.[13]

As if this close encounter with death and destruction wasn't enough for one day, away on a nearby stretch of the Missouri, one of the boats carrying the journals, navigating instruments, and medicine of the crew was almost capsized by a sudden squall. Fortunately, while Lewis and Clark watched helplessly from the bank, the boat rolled on its side before the crew gathered their wits, took in the sail, and managed to steer the vessel into the wind. With the turbulent current flooding into the boat, precious supplies and journals began floating out. But Sacagawea exhibited cool countenance in the face of imminent tragedy and calmly gathered up all the articles she could as the crew rowed the water-logged pirogue to the bank.

Chronicling the accident that night in his journal, Lewis admitted that the expedition

had been halted by an occurrence, which I have now to recappitulate, and which altho' happily passed without ruinous injury, I cannot recollect but with the utmost trepidation and horror.[14]

They spent most of the next two days camped on the opposite bank of the Missouri, a few miles upstream from the mouth of Snow Creek, whose headwaters originate very close to the Engdahl ranch where we camped when I was a student at Berkeley. Similarly to Medea, Sacagawea had saved the expedition. A still shaken Lewis, writing two days later on the 16th, after the crew had evaluated their losses and dried out all the items that could be salvaged, recorded that

> the loss we sustained was not so great as we had at first apprehended; our medicine sustained the greatest injury, several articles of which were intirely spoiled, and many others considerably injured, the balance of our losses consisted of some gardin seeds, a small quantity of gunpowder, and a few culinary articles which fell overboard and sunk. The Indian woman to whom I ascribe equal fortitude and resolution, with any person onboard at the time of the accedent, caught and preserved most of the light articles which were washed overboard. . . . [15]

It was sobering to note how tenuously the expedition's "thread of life" was strung by the Fates, as they struggled through the breaks of Hell Creek. Their bodies and minds were certainly being tempered by those badlands in ways that they had never before experienced. Without much imagination, it was easy to envision how the Missouri could have quickly been transformed into the Acheron, with Charon rowing the unlucky crew to the gates of Hades.

Yet another adrenaline rush befell Clark on May 17, as the

Corps progressed past today's Seven Blackfoot Creek. This stream is near the western edge of the Hell Creek region, and its ravines and gorges constituted an occasional collecting area for our crews from Berkeley. While walking along the southern bank of the Missouri, Clark inadvertently wandered too near a rattlesnake, which launched a defensive strike that nearly nicked him. Most appropriately, Seven Blackfoot Creek was originally named Rattlesnake Creek by Clark and the Corps. Despite the potentially perilous encounter, Lewis recorded the event in a rather detached scientific description:

> Capt. Clark narrowly escaped being bitten by a rattlesnake in the course of his walk, the party killed one this evening at our encampment, which he informed me was similar to that he had seen; this snake is smaller than those common to the middle Atlantic States, being about 2 feet 6 inches long; it is of a yellowish brown colour on the back and sides, variagated with one row of oval spots of a dark brown colour lying transversely over the back from the neck to the tail, and two other rows of circular spots of the same colour which garnis the sides along the edge of the scuta. It's bely contains 176 (s)cuta on the bely and 17 on the tail.[16]

This detailed anatomical litany represents the first scientific description of the prairie rattlesnake, which represented a new subspecies later named *Crotalus viridus viridus*.

I can attest that many of the marvels that intrigued the Corps still dwell around Hell Creek. Just as Lewis and Clark witnessed, coyotes and deer still dart through the badlands and pronghorn still bound on the prairie. Other intimidating animals, such as the rattlesnake that struck at Clark, still slink across the slopes. I have sensed the same rush of adrenaline he must have felt as I leapt away from an unexpected encounter with one of their modern-day descendants. But fortunately for fossil collectors like me, brown bear and bison have been banished from the Breaks.

After setting camp upstream from its mouth on the Missouri, their night's rest was rudely interrupted when the night guard sounded the alarm by warning

> of the danger we were in from a large tree that had taken fire and which leant immediately over our lodge. We had the lodge removed, and a few minutes after a large proportion of the top of the tree fell on the place where the lodge had stood; had we been a few minutes later we should have been crushed to attoms.[17]

Over the next two days, they continued up the river, occasionally gaining vantage points where they could see the Little Rockies in the distance. By May 20 they passed present-day Squaw Creek, which they named "Blowing-fly Creek"

> from the emence quantities of those insects which geather on our meat in such numbers that we are oblige to brush them off what we eate.[18]

By 11 A.M. they had reached the mouth of what they named the "Muscel Shell River," now only slightly altered to the Musselshell. As Clark recounts, they were

> 2,270 miles up [from the mouth of the Missouri] ... I measured it and find it to be 110 yards wide, the water of a Greenish Yellow Colour, and appers to be navagable for Small craft.[19]

The hard labors of their mountain crossing lay just ahead, but they had successfully transited through the perils at Hell Creek. The next five and a half months saw the Corps labor determinedly up the eastern flank of the Rocky Mountains, and cross the continental divide at Lemhi Pass on the modern-day border between Montana and Idaho. Already exhausted and fearful of the approaching freeze, they descended the western flank of the Rockies and floated down the Snake and Columbia Rivers to the Pacific Coast, near today's Astoria, Oregon. There they spent the exceedingly wet winter months of 1805–06 before heading back east. In early July 1806, near what is now Lolo, Montana, Lewis and Clark split their crew to explore disparate domains east of the continental divide. While Clark embarked down unexplored reaches of the Yellowstone River, Lewis started back down the Missouri. His contingent spent eleven days from July 17 to July 28 exploring the drainage of what the Corps christened the Marias River, in honor of Lewis's cousin, Maria Wood. From that point Lewis hastened down the Missouri and back through the Hell

Creek region, passing the mouth of the Musselshell on August first. Although rain forced them to spend the second in camp drying their gear, by the fourth they had passed the mouths of Snow Creek, Hell Creek, and the Big Dry, which was now flowing sixty yards wide. They were anxious to get to the mouth of the Yellowstone, where they had planned to rendezvous with Clark and his detachment.

Despite having twice traversed one of the richest paleontological field areas in the world, Lewis and Clark did not record any discoveries of dinosaur bones in the immediate area of Hell Creek. This is probably because, as I mentioned in the previous chapter, the dinosaur-bearing sediments are not located deep in the Breaks, where the Corps passed along the banks of the Missouri, but higher up on the bluffs. More puzzling is why they did not record the presence of plentiful ammonite fossils, with their prominent pearly shells, that are strewn throughout the Bearpaw Shale along the banks of the Missouri. Perhaps they simply did not recognize them as fossils.

Nonetheless, Clark did collect one fossil on his return trip along the Yellowstone River near Pompey's Pillar, just east of the modern city of Billings. Mistaking the bone for that of a fish, Clark recorded the following account in his journal on July 25, 1806:

Dureing the time the men were getting the two big horns which I had killed to the river I employed my self in getting pieces of the rib of a fish which was Semented within the face of the rock this rib is about 3 inches in Secumpherence about the middle it is 3 feet in length tho a part of the end appears to have been broken off. . . . I have several pieces of this rib[20]

Unfortunately, the specimen is now lost, so its true identity cannot be confirmed, but based on the size and subsequent research in the rocks of that specific locale, there is not much doubt that Clark had found the first dinosaur fossil from the American West. His discovery set the stage for extensive paleontologic prospecting at Hell Creek one hundred years later. But before that could happen a century of settlement in the northern Great Plains would trigger decades of discord with the Native Americans, who, for the most part, had played such a pivotal role in the Corps' success. These conflicts between the Native Americans and the United States Army in support of the settlers from the East would punctuate and drastically alter the course of biologic and social evolution in the West, as well as the shape of history in the twentieth century. Once again, Hell Creek would serve as the setting for crucial actors to perform their pivotal roles.

4

IN THE WAKE OF CUSTER'S RUIN

Lewis and Clark were obviously not the first humans to explore and settle the upper drainage of the Missouri River. At the time of their expedition, it was well known that numerous tribes of Native Americans inhabited the broad stretches of prairie and deeply etched breaks on both the North and South sides of the river. Throughout the year these nomadic tribes hunted bison and other game, throughout loosely defined and often overlapping territories with other distantly related kin.

Before Lewis left Washington, D.C., as Ambrose relates:

Jefferson and Lewis talked at length about these tribes, on the basis of near complete ignorance. They speculated that the lost tribe of Israel could be out there on the Plains, but it was more likely, in their minds, that the Mandans were a wandering tribe of Welshmen.[1]

Thomas Jefferson gave Lewis very specific instructions concerning his interactions with these indigenous inhabitants of the Upper Louisiana Territory. Ambrose recounts[2] that their unusual speculations concerning the heritage of these tribes rendered Jefferson's guidelines to Lewis for interacting with the Native Americans "hopelessly naïve and impossible to carry out."

Nonetheless, Jefferson was intent on trying to establish peaceful relations that would benefit both the Indian tribes and the immigrants from the East Coast who he envisioned would soon migrate west in order to reap the resources that he sent the Corps of Discovery to inventory.[3] Jefferson instructed Lewis to carry a message to each tribe that the United States sought to integrate them into a mutually beneficial system of commerce founded on peaceful relations. The establishment of this trading network would require the tribes to abandon their weapons and take up trapping. Although Jefferson was aware that the Sioux or another tribe of warriors might react unfavorably and attack the expedition, he demanded that Lewis do everything in his power to avoid armed conflict. Nonetheless, if push inescapably came to shove, Lewis's highest priority was to successfully cross the continent, even if that meant protecting the expedition through use of force.

Lewis abided by Jefferson's orders quite closely. Only once during the whole expedition did he feel obliged to shoot a Native American in response to an aggressive action taken against the crew. That incident occurred on Lewis's return trip down the Missouri, when his detachment of four men encountered a small party of eight Blackfeet during their reconnaissance of the Marias River. After the young warriors tried to steal rifles and horses from the crew, one was stabbed to death and one was shot in the stomach during the ensuing melee. Although such skirmishes were rare, Jefferson, through Lewis, had established a precedent of sorts with the Native Americans of the region, and the ensuing conflicts would bedevil the United States government for most of the rest of the century. The nadir of these policies occurred soon after the Civil War, as the country refocused its efforts to settle the lands of the Dakota, Montana, and Wyoming Territories. Basically, the plan involved the same tribes of Sioux that worried Jefferson and Lewis.

As Jerome Greene outlines in *Yellowstone Command*,[4] the Teton, or Lakota, Sioux sought to evade conflict with the more potently armed Chippewas and to expand their territorial boundaries in the last few decades of the 1700s. They migrated west from the area around the Great Lakes to occupy the region from the Yellowstone on the north to the Platte on the south and from the Missouri on the east to the Bighorn on the west. These were prime plains for hunting bison, and many of the neighboring tribes who previously occupied the area were anything but pleased. Most acutely affronted were the Crows, who were quickly displaced by the Sioux and their Cheyenne allies, catalyzing decades of vengeful skirmishes and raids.

Beyond that, the Teton Sioux and the Northern Cheyenne inevitably formed a formidable barrier to the western expansion of the American settlers from the east. Clashes with the U.S. Army erupted in 1854 when the Tetons obliterated a small detachment near Fort Laramie in Wyoming, followed by a vengeful massacre of Sioux noncombatants by the army at Ash Hollow, Nebraska, thereby triggering several decades of armed conflict. What turned out to be a temporary lull followed the signing of a treaty four years after the conclusion of the Civil War.

As Greene explains, the Native Americans grudgingly signed the second Treaty of Fort Laramie with the U.S. government in 1868, after two years of hostilities between the army and the Sioux. This treaty stipulated that the Indians must live primarily "within prescribed boundaries approximating those of the present state of South Dakota west of the Missouri River." However, the government did grant the tribes the right to hunt in the "Wyoming Territory west to the Big Horn Mountains and south to the North Platte River."[5]

As originally envisioned, the Sioux retained exclusive rights to live and hunt in the Black Hills region of present-day South Dakota—a sacred land, bounteous in game and forests, that Sioux legend identified as the tribe's cradle of origin. Yet, the Sioux were not restricted to this "Great Sioux Reservation." They also held a right to unrestricted movement within the "unceded" plains of the Wyoming Territory west to the Bighorn Mountains.

The first signs of conflict did not take long to surface after this stopgap treaty was signed. The Northern Pacific Railway set it sights on the valley of the Yellowstone River in 1871 and 1872. Military escorts under the command of a thirty-four-year-old lieutenant colonel, George Armstrong Custer, were sent to protect the crews. At the time, Custer was already a highly decorated Civil War commander, who had been awarded an honorary rank of brigadier general. During that war, he had carefully cultivated not only his unquestioned capacity for valor but also his visual image on the battlefield. As Evan Connell notes in *Son of the Morning Star*:

> While he served on McClellan's staff his natural exuberance blossomed into what could only be called flamboyance. . . . He began to wear a tightly fitted hussar jacket, gold lace on his pants, and rebel boots. One staff member likened him to a circus rider.[6]

Never missing an opportunity to rub salt in the wounds of his adversaries, Custer concocted an eye-catching cameo for his appearance at the surrender ceremony of his Confederate opponent, General Joseph Kershaw. As related by Connell in Kershaw's own words.

> A spare, lithe, sinewy figure; bright, dark, quick-moving blue eyes; florid complexion, light wavy curls, high cheekbones, firm-set teeth—a jaunty close-fitting cavalry jacket, large top-boots, Spanish spurs, golden aiguillettes . . . a

quick nervous movement, an air telling the habit of command—announced the redoubtable Custer. . . .[7]

Since 1866, after the conclusion of the Civil War, he had led the Seventh Calvary on numerous campaigns against the Indians west of the Mississippi. Custer continued to be a lightning rod for his colleagues' emotions. As Robert Utley states, throughout his life and even after his death Custer's flamboyance almost always triggered either unconditional loyalty or unbridled animosity from his cohorts:

Some saw him as reckless, brutal, egotistical, selfish, unprincipled and immature. Others looked upon him as

Portrait of the legendary Lakota Sioux Chief Sitting Bull. (NATIONAL ANTHROPOLOGICAL ARCHIVES, SMITHSONIAN INSTITUTION 3195-G)

upright, sincere, compassionate, honorable, tender and above all fearless in battle and brilliant in leading men to victory.[8]

The treaty of 1868 specifically allowed the building of railroads through the unceded territory of the Yellowstone region, but Sioux warriors under their legendary Chief Sitting Bull registered their discontent at this perceived invasion, launching a series of determined attacks.

Then, once rumors of gold discoveries in the Black Hills of present-day South Dakota began to seep out of the region, all bets were off. The army sent Custer, along with his troops of the Seventh Cavalry and two miners, to investigate whether gold actually could be found in the region, although the "official" reason for the foray was to scout for the site for a fort.

On July 23 the miners informed Custer that they had indeed found gold dust and nuggets the size of pinheads in the bed of a stream nearby French Creek. As Connell states:

> At that moment the climactic battle became certain. The insistent rumors of half a century were verified.[9]

Although the army tried halfheartedly to keep prospectors out of the Black Hills and the Great Sioux Reservation, in reality President Grant and his senior military officers agreed, in effect, to look the other way and let the prospectors do as they

The major battles of the Great Sioux War, 1876–1877. (REDRAWN WITH PERMISSION FROM JEROME A. GREENE, *YELLOWSTONE COMMAND*, 1991, P. 17.)

pleased. By early 1875 the rush was on, and by May of the following year reports of Indians ambushing the white fortune
seekers were printed in local newspapers. In order to try and
control the Sioux and other tribes still roaming outside the
reservations, the government issued orders that all the Indians
would have to surrender to government authorities by January
31, 1876, or hostilities would follow. No one really expected
Sitting Bull and the other Sioux chiefs to do so, and in effect
the ultimatum was simply an order that, if violated, would give
the Army an excuse to attack. Sitting Bull and his cohorts
ignored the government's directive, and the deadline passed
without any meaningful response.

The inevitable and pivotal army campaigns related to the
Great Sioux War were initiated in early 1876. Three columns
of troops totaling 2,500 men were assembled by the famous
Civil War General of the Army, William Tecumseh Sherman,
and the commander of the Missouri Division, Gen. Philip
Sheridan. One, under the command of Brig. Gen. George
Crook, originated from the Army's Department of the Platte to
the south in Wyoming. Another, commanded by Col. John
Gibbon, traveled from the west near Bozeman, and the third,
led by Brig. Gen. Alfred Terry, marched from the Dakota Territory to the east. Their mission was to confront the Sioux in
their summer hunting grounds just south of the Yellowstone in
the drainages of the Powder and Bighorn Rivers. At the start of
the summer, although estimates vary widely as to the number
of tribesmen in the conglomeration of the Indian camp, Utley

suggests that about three thousand were present in total, including roughly eight hundred warriors. The column coming up from the south under the command of Brig. Gen. George Crook first encountered the warriors during a coffee break on the morning of June 17, in the upper reaches of Rosebud Creek. Somewhere between seven hundred and fifty and one thousand Sioux and Cheyenne warriors, led by the steely and mysterious Oglala chief, Crazy Horse, ambushed Crook's column.[10] The battle lasted for six hours, with combatants on each side fighting courageously, before the Indians withdrew from the field and Crook's battered force attended their wounded.

In haste, Crook retreated until he could connect with reinforcements, spending the next ten days licking his wounds and fishing for the abundant trout in nearby streams. Although more Native Americans than soldiers were killed, this Battle of the Rosebud set the stage for the climax to come. Although historians still debate which side won or lost, Crook's encounter and retreat along the Rosebud assured that the troops coming in from the north would lack the essential support they needed from Crook's troops to the south.

Meanwhile, the columns of troops originating from Dakota and Bozeman were converging near the mouth of the Rosebud along the Yellowstone to the north. They were completely unaware of Crook's exact position, as well as his frightful engagement. Embedded within the Dakota column under the command of Brig. Gen. Alfred Terry were several companies of the Seventh Calvary led by the seemingly ubiquitous Custer, impatiently itching for another fight. On June 22 Custer paraded his troops before General Terry before embarking up Rosebud Creek. The elements comprised 597 officers and enlisted men, 35 Indian scouts, including Arikaras, Crows, and even a few Sioux, along with a dozen or so other civilian guides and packers. In order to make sure none of his heroic exploits would go unrecorded, Custer even invited a newspaper journalist along.

The general plan devised by Terry and Gibbon, in consultation with Custer, was for Custer to swing south along the Rosebud before turning west into the valley of the Little Bighorn, making sure that the Indians did not try to escape towards the east or south. Meantime, Terry and Gibbon would go up the Bighorn from the Yellowstone to the mouth of the Little Bighorn, to keep the Indians from escaping to the north and to catch them between their forces and Custer's. Not knowing the exact location of the Indian village, Terry intentionally gave Custer a good deal of leeway for improvisation. As Terry's orders to Custer stated:

> It is, of course impossible to give you any definite instructions in regard to this movement, and were it not impossible to do so the Department Commander places too much confidence in your zeal, energy, and ability to wish to impose upon you precise orders which might hamper your action when nearly in contact with the enemy.[11]

It seems pretty certain that Terry expected Custer to "make the kill" with his more mobile forces, while Terry and Gibbon would seek to block any attempted escape. But Gibbon was concerned about Custer, later writing Terry:

So great was my fear that Custer's zeal would carry him forward too rapidly, that the last thing I said to him when bidding him good-by . . . was, 'Now, Custer, don't be greedy, but wait for us.' He replied gaily as, with a wave of his hand, he dashed off . . . [12]

A day later the scouts had located the Indian trail leading west into the valley of the Little Bighorn. As they moved closer the scouts perceived troubling signs, and they began to realize the enormous size to which the village had grown. In fact, over the last several days, numerous Indians from the regional reservations had flocked to Sitting Bull's camp for the summer hunting and rituals. Now, nearly one thousand lodges and seven thousand Indians had gathered along the Little Bighorn, including two thousand warriors. Five separate tribal circles contained the Sioux tribes, including Hunkpapas, Oglalas, Miniconjous, Sans Arcs, Blackfoot, Two Kettles, Brules, and other minor groupings. Another sizable tribal circle contained the Cheyennes. The list of leaders included most of the preeminent chiefs of the day: Gall, Crow King, and Rain-in-the-Face represented the Hunkpapas; Crazy Horse and Low Dog the Oglalas; Red Horse the Miniconjous; Spotted Eagle the Sans Arcs; Jumping Bear the Blackfoot; and Two Moon the Northern Cheyenne. But over all towered the legendary Sitting Bull. In addition to outstanding courage as a warrior in his early years, he had developed other formidable influences over the tribes in spiritual and political matters. As Connell relates, a Canadian correspondent provided a vivid description of Sitting Bull when he encountered the chief after he had fled north of the border after the battle:

> An Indian mounted on a cream-colored pony, and holding in his hand an eagle's wing, which did duty for a fan, stared solidly, for a minute or so, at me. His hair, parted in the ordinary Sioux fashion, was without a plume. His broad face, with a prominent hooked nose and wide jaws, was destitute of paint. His fierce, half-bloodshot eyes gleamed from under brows which displayed large perceptive organs, and, as he sat there on his horse regarding me with a look which seemed blended of curiosity and insolence, I did not need to be told that he was Sitting Bull.[13]

Connell also explains that, although his name might not sound impressive to us, as it is translated in English, in Sioux it carried a slightly different meaning. First, although whites considered the bison to be egregiously unintelligent, the Native Amricans regarded the animal as the wisest and most powerful creature next to their Great Spirit. In addition, the word "sitting" in Sioux carries an implication of residing, so Sitting Bull "signified a wise and powerful being who had taken up residence among them."

Both the Native Americans and the U.S. Army experienced moments of savage victory and humiliating defeat in the Great Sioux War. But this would be the last great moment of exhilaration for the Native Americans. The stage was set for a conflict that would have massive repercussions for both sides.

As the Seventh Cavalry neared the Indian encampment from the southeast during midday on June 25, Custer divided his troops into three battalions. One, comprised of 125 men under the command of Capt. Frederick Benteen, swung out to the southwest in order to ensure that the Indians did not escape to the south down the drainage of the Little Bighorn. Another, containing 140 soldiers under Maj. Marcus Reno, moved directly down a small drainage to the valley of the Little Bighorn and mounted an attack on the Indians from the south. Although the Indians were aware that Custer's force was in the general area, the attack itself came with little warning. Nonetheless, it was quickly rebuffed by an overpowering counterattack led by Sioux warriors, and Reno lead a disorganized retreat to a bluff on the east side of the river opposite the Indian encampment. Shortly after, Benteen's contingent would join with Reno's and wage a desperate defensive action that would last through the night and into the following afternoon.

Meanwhile, Custer lead a squadron of 225 soldiers along the bluffs on the east side of the river in an apparent attempt to outflank the village and attack from the east or northeast. As Connell documents, Custer enthusiastically waved his hat to spur on his troops as the four-mile-long village finally came into view. Later, Benteen's trumpeter reported that Custer exclaimed, "Hurrah, boys, we've got them!" Connell concludes that "If indeed Custer made such a remark after sighting the greatest concentration of militant Indians in the history of North America it sounds like a joke from an old vaudeville routine."[14]

With Reno pinned down on the bluffs, many warriors were freed up to confront Custer at nearly full force. As part of Custer's contingent attempted to cross the Little Bighorn and attack the encampment down a coulee from the east, they were initially repelled by a small band of Sioux. But within minutes Gall rallied a massive force that crossed the river from the west and chased all of Custer's battalion to the north along the bluffs. This served to isolate Custer from his other commanders and thwart any attempts to consolidate their forces. As Custer moved farther north along the bluffs, apparently seeking a place to attack from the northeast, he was quickly confronted by another massive party of Indians led by the legendary war chief Crazy Horse. The tables had been turned, for, instead of Custer catching the Indians in a vise between his forces and Reno's, the Indians now had Custer caught between the forces led by Crazy Horse and Gall. At this point the rout was all but inevitable.

Estimates vary concerning the number of soldiers in Custer's contingent who were killed, but none survived. Utley states that 210 of Custer's contingent were massacred, while 53 of Reno and Benteen's men were killed while retreating and defending their tenuous position on the southern bluffs. Although Custer himself was said to have suffered only bullet wounds to the chest and his left temple, many of Custer's dead had been more severely mutilated, with slashes on various parts of their bodies, scalps taken and even limbs and other body parts severed. In part, these mutilations represented signs left by the warriors to identify the specific tribe responsible. But squaws in their grief for lost loved ones also took part.

As Connell relates, this ritualistic disfigurement performed by the Indians may simply reflect their "grief and bewilderment." In essence they wondered, why had the army tracked and attacked them, "when all they ever wanted was to be left alone so that they might live as they had for centuries: hunting, fishing, trailing the munificent buffalo."[15] They simply wished to continue their nomadic lifestyle and had no interest in settling down on reservations as farmers. The concept of land ownership was completely foreign to them, since they believed that the land belonged to all people. Thus, no person or group had the right to claim and clear the land by plowing and killing the plants that their god had put there. Also playing a role in the mutilations was the belief among some Native Americans that severed body parts could be of no use to the enemy in the afterlife, thereby rendering the victim helpless in any future conflict.

Exactly what happened during "Custer's last stand" is more a matter of conjecture than fact. The troops under the command of Reno and Benteen, pinned on the bluffs several miles to the south, did not have a direct line of sight and were too far away to see. None of Custer's contingent who actually

participated in the action survived, and accounts by the warriors vary widely.

What is known is that during the evening of June 26 the
enormous Indian encampment packed up and moved to the
south toward the Bighorn Mountains, presumably alerted by
their scouts that a column of troops was approaching their
position from the north. That column contained the troops led
by Terry and Gibbon, and a couple of officers from Reno's command rode out to meet them, asking in honest consternation,
"Where is Custer?" The commander of Gibbon's scouts had
already discovered the carnage, and the answer was clear. As
Utley writes:

> On the morning of June 28, Reno's men rode to the
> Custer battlefield to bury the dead. "A scene of sickening
> ghastly horror," Lieutenant Godfrey recalled . . .

> The dead were buried hastily. Tools were few, and in
> most cases the burial details simply scooped out a shal-
> low grave and covered the body with a thin layer of sandy
> soil and some clumps of sagebrush. The officers were
> buried more securely and the graves plainly marked for
> future identification.[16]

At the beginning of July 1876 the country was caught up in
preparation for a massive party—the centennial celebration of
Jefferson's Declaration of Independence. By July 3 messengers

from Terry and Gibbon's column had reached Bozeman with the news, and by the fourth and fifth most of the country was aware of the slaughter. Reactions generally ranged from dumbfounded shock to shrill calls for retribution. In addition, the country was gearing up for a presidential election that fall, and some argue that one reason Custer forced the attack without waiting for Terry and Gibbon to support him was because he had aspirations to run for that office. In any event, the United States was hell-bent on revenge.

Yet, rather than confront the massive force pursuing it, the amalgamation of tribes responsible for Custer's defeat were not interested in further engagements. Utley sums up their motivations in his book.[17] In addition to the threat posed by Terry's advancing column, the sheer size of the Indians' encampment created numerous logistical problems. Most of the game in the area had already fled, leaving little for the warriors to hunt and the tribes, as a whole, to eat. Within days after leaving the battlefield, small tribal contingents left the larger assembly to head off on their own or return to the reservation. Meantime, the remnants of the larger encampment traversed the divide between the Little Bighorn and the Rosebud. After following that drainage downstream for a while, they swiftly headed east toward the Tongue River, hunting for bison and covering their tracks by igniting grassfires as they went.

Terry, Gibbon, and Crook were completely intimidated by the enormity of the massacre and decided to wait for reinforcements before pressing their pursuit. It was August before reinforcements arrived. In the meantime the Indian force fragmented, with some even returning to reservations. Sitting Bull guided his followers northeast toward the valley of the Little Missouri River, while Crazy Horse led his tribe south toward the Black Hills. After Terry and Crook split their forces in early September, Crook did encounter a small band of Crazy Horse's contingent near Slim Buttes just across the border in the Dakota Territory. A skirmish ensued in which the small band of Oglalas was decimated and a counterattack by Crazy Horse's warriors was fought off. Weary and dangerously low on supplies, all the combatants withdrew.

But what does the Battle of the Little Bighorn have to do with the area around Hell Creek? In fact, the linkage is quite lucid. Within a week after news of Custer's defeat had reached Fort Leavenworth in Kansas, a gritty, thirty-six-year-old commander prepared his troops of the Fifth Infantry to move out to the front of the Great Sioux War along the Yellowstone River. Col. Nelson A. Miles not only counted Custer as a friend, he was "a practitioner of [Custer's] aggressive, hard-hitting style of war," according to Utley.[18] As Jerome Greene documents in *Yellowstone Command*, Miles had utilized that style to considerable success against the Kiowas, Comanches, and Southern Cheyennes during the Red River War of 1874–75, and had established the Fifth as one of the nation's premier combat units in encounters with the Native Americans. Miles's contingent was part of the reinforcements sent by the army to beef up the force in the wake of Custer's defeat.

Portrait of Colonel Nelson A. Miles, taken by photographer Stanley J. Morrow in 1877.
(PRINTED WITH PERMISSION FROM THE MONTANA HISTORICAL SOCIETY)

They arrived in the vicinity of the Powder and Tongue Rivers on August 1 and continued up the Yellowstone to Terry's camp near the mouth of the Rosebud, arriving on August 2. A few skirmishes with small groups of Indians along the Yellowstone suggested that some of the tribal elements were trying to move north across the river. After a few days during which Miles's contingent accompanied Terry's troops south along the valley of the Rosebud, Terry's column met Crook's troops coming up from the south. Within hours after that reunion, Miles was directed to return to the Yellowstone and patrol the river in an attempt to keep the Sioux and Cheyenne from crossing. Although frustrated by the lack of action, Miles was heartened by gaining his independence from

Terry and Crook. As Greene relates, Miles was not impressed with their courage or command structure. As he wrote his wife.

> I never saw a command so completely stampeded as this ... (Terry) does not seem very enthusiastic or to have much heart in the enterprise.[19]

Other passages related by Greene reveal that Miles had criticized the leadership of Terry and Crook well before he had joined their command:

> I think it almost a military crime ... that these two commands are not under one head and governed by the simplest principles of warfare ... One little success would change the whole feeling in regard to those Indians.[20]

And Miles was convinced that he was the man to do it. Upon reaching the Yellowstone on August 11 he aggressively set about deploying his forces at likely fording spots as far downstream as O'Fallon's Creek past the Powder River. Sightings and occasional firefights established that small groups of Indians were indeed attempting to cross, with one group of several hundred apparently crossing soon after the soldiers passed just below the mouth of the Tongue.

Miles and his contingent spent most of September and the first part of October constructing a cantonment at the mouth of the Tongue River. Supplies for the base were transported up the Yellowstone by steamers and wagon trains, which also served to help keep track of the tribes' movements. Based on a plan approved by Sheridan and enthusiastically endorsed by Miles, he would use the small settlement on the Tongue as a base of operations to maintain a permanent presence through the winter in the Indians' traditional hunting grounds. Then he would pursue them at a time when conditions inhibited their mobility. The winter would most certainly be brutal, with snows, strong winds, and temperatures dropping well below freezing. Yet, as he later recalled:

My opinion was that the only way to make the country tenable for us was to render it untenable for the Indians . . . I was satisfied that if the Indians could live in that country in skin tents in winter, . . . we, with all our better appliances could be so equipped as to not only exist in tents, but also to move under all circumstances.[21]

Miles began accumulating and storing winter food and clothing for his men. He also recruited a group of scouts from both the friendly Crows and frontier trappers, such as Luther "Yellowstone" Kelly. In small mobile groups, the scouts kept an eye on the movements of the Native Americans in the region. As Greene writes, the army's intelligence suggested in September that Sitting Bull and his entourage roamed across the region in Dakota near Twin Buttes, along the Grand River, before heading toward the bison-rich prairies above the lower reaches of the Yellowstone. Quite simply, he sought ample game, and, by all accounts, neither Sitting Bull nor Crazy Horse desired to confront the army.[22]

Sitting Bull had long ago fallen in love with the landscape just south of Hell Creek, which consistently teemed with bison. Although that land had been designated as belonging to the Assiniboin tribe in the 1851 Treaty of Fort Laramie, such formalities had little effect on the great Sioux leader. In defending his privileges to hunt in the region, he had long been willing to take on all comers. As Jack McRae relates in *Trailin' Through Time*,[23] a low but intricately sculpted outcropping of yellowish-gray sandstone, called Crow Rock, lies in the southeast corner of Garfield County, which served as the focal point for a siege that pitted Sitting Bull's warriors against a party of Crow braves. As documented by Stanley Vestal in *Sitting Bull, Champion of the Sioux*, the incident occurred in the autumn of 1869, when a band of about thirty Crow encountered two young Sioux, belonging to Siting Bull's village, on a bison hunt. Although the Crow killed one of the Sioux, the survivor evaded capture and fled back to Sitting Bull's encampment. Enraged, Sitting Bull set out with one hundred Sioux warriors in pursuit of the Crow murderers, tracking them along the ridge that separates Crooked Creek from Custer Creek, before crossing the divide between the Yellowstone and Missouri. After being pursued down the drainages of Hay Creek and Crow Rock Creek, the Crow holed up in the center of the protective outcroppings of Crow Rock. The climatic "battle was intense but short. All the Crow were killed, fourteen of the Sioux died and eighteen were wounded."

Just seven years after the incident at Crow Rock and just four months after annihilating Custer, as Sitting Bull and his contingent arrived in the area north of the Yellowstone just south of the area around Hell Creek, they were annoyed to find troops stationed there, running wagon trains and steamers along the river. On October 10 a supply train started upriver for the Tongue from the cantonment at Glendive accompanied by about 160 soldiers. That night and into the following day hundreds of warriors harassed the train from a safe distance with rifle fire. The train finally reversed course and returned to Glendive. The attacking party was allied with Sitting Bull, and led by Hunkpapa, Sans Arc, and Minneconjou chiefs such as Gall, Bull Eagle, No Neck, Pretty Bear, and Red Skirt.

As Greene relates, the tribesmen planned on enjoying their yearly bison hunt before continuing north to Fort Peck to trade.[24]

This course would take them directly through the eastern portion of the Hell Creek region.

Undaunted, the commander at Glendive, Lt. Col. Elwell Otis, assembled a larger force of 185 men and launched another train on October 14, in a second attempt to reach Miles. But again the supply train met resistance on the bluffs and creeks that drained south into the Yellowstone along the northern bank. On the sixteenth, near Cedar Creek, as Greene reports, a warrior approached Otis and his detachment, driving a stake with a note on white cloth into the summit of a nearby hill. Once the brave fled an army scout rode out and brought the note back to Otis, who discovered this message:

YELLOWSTONE.
I want to know what you are doing traveling on this road. You scare all the buffalo away. I want to hunt on the place. I want you to turn back from here. If you don't I will fight you again. I want you to leave what you have got here, and turn back from here.

I am your friend,
SITTING BULL.
I mean all the rations you have got and some powder. Wish you would write as soon as you can.[25]

Otis responded through his scout that he intended to continue "and that we should be pleased to accommodate them any time with a fight." After more discussions and threats, the Indians withdrew, having been denied ammunition, but supplied with two sides of bacon and 150 pounds of bread.

Meanwhile, Miles, worried about why the train had not arrived on time, rode out to investigate with a large force of 434 soldiers. Meeting Otis on the eighteenth after the confrontation, he determined to pursue Sitting Bull's forces. On the morning of the twentieth, after Miles's contingent marched about twenty miles northeast toward the divide between the Yellowstone and Hell Creek area, several hundred warriors confronted Miles's troops on nearby hills and ridges near Cedar Creek. Messengers informed Miles that Sitting Bull wanted to talk. After several hours of preparations the two men and their entourages met tensely between the forces arrayed against each other on a buffalo robe spread out by Sitting Bull. Greene paints a portrait of the two adversaries. Miles's heavy uniform featured a fur cap and a long overcoat adorned with bear fur on the collar and cuffs. From then on Miles was referred to as the "Man with a Bear Coat" by the Native Americans. Sitting Bull's outfit, on the other hand, consisted simply of "leggings, moccasins and a breechcloth wrapped about with a buffalo robe." His other common accoutrements, such as feathers, were missing.[26]

In violation of their previously negotiated agreement, numerous weapons were concealed among the members of both entourages. Miles noted his impressions of the imposing Hunkpapa chief:

> He was a strong, hardy, sturdy looking man of about five feet eleven inches in height, well-built, with strongly marked features, high cheek bones, prominent nose, straight, thin lips, and strong under jaw, indicating determination and force. He had a wide, large, well-developed head and low forehead. He was a man of few words and cautious in his expressions, evidently thinking twice before speaking. He was very deliberate in his movements and somewhat reserved in his manner. At first he

was courteous, but evidently void of any genuine respect for the white race.[27]

Basically, Sitting Bull wanted Miles to leave him alone. He also desired to hunt bison and trade for ammunition. In return, he would promise not to attack the soldiers. Miles insisted that all of Sitting Bull's cohort surrender and report to reservations. Greene reports that "Miles angered Sitting Bull by telling him that he had learned of his intended movement to the Big Dry River to hunt buffalo," an area which forms the southern and eastern margins of the Hell Creek region. The council ended inconclusively but peacefully, though each leader suspected that the other intended to kill him.

Looking northeast at the low bluffs and ravines along the drainage of Cedar Creek northwest of Terry, Montana. This is how the battlefield, where Miles first confronted Sitting Bull in October 1876, appears today. (LOWELL DINGUS)

The next day a similar council proved fruitless, and the leaders returned to their forces with entourages in tow. Hostilities soon followed, as warriors set the grass afire in coulees near the troops. By evening Sitting Bull and his cohort had taken flight to the south down Bad Route Creek toward the Yellowstone, setting more fires as they went. Miles pursued them for forty-five miles over the smoldering prairie, reaching the Yellowstone on the twenty-third after most of the Native Americans had crossed near the mouth of Cabin Creek.

Another council was arranged on the twenty-fifth, and, although Gall attended, Sitting Bull did not. Miles discovered that, during the pursuit, Sitting Bull had doubled back north toward the Big Dry and Hell Creek to continue hunting. It was not the last time that Miles would be outfoxed. As Greene states, Sitting Bull's entourage would soon swell to almost four hundred people, as the followers of Gall, Pretty Bear, and other chiefs joined the village.[28]

Some of the chiefs Miles chased to the Yellowstone agreed to surrender and return to their reservations, although not many followed through; but Miles's main quarry remained at large. Sitting Bull and the bands with Four Horns, Black Moon, and Iron Dog crossed the divide between the Yellowstone and Missouri, traveling right through the eastern part of the Hell Creek region.

Miles returned to his Tongue River base to regroup, and received word on November 2 that Sitting Bull's band was camped on the Big Dry, just twenty miles south of the Missouri, along the eastern edge of our paleontological field area. Characteristic of the region's seasons, temperatures were already plummeting below zero, with snow and wind chills as low as sixty degrees below zero. Troops fashioned overcoats, leggings, hats, gloves, and overshoes from bison hides in an attempt to withstand the onslaught.

Despite the hostile weather, Miles remained undeterred, setting out on November 6 with 434 soldiers in pursuit of Sitting Bull. Greene describes their tortuous trek through the Hell Creek region,[29] as the troops marched almost twenty miles and camped near the Big Dry on November 9. Scouts spotted a fresh Indian trail on the tenth, and Miles suspected that he was hot in pursuit of Iron Dog's village, which then totaled 119 lodges. But traipsing though the drainage of the Big Dry proved both tedious and treacherous, as the wagons and pack animals sometimes broke though the sheet of ice covering the river. One soldier recorded that "The country on either side is very broken, much of it being bad land, and high, precipitous, sterile bluffs . . . During the night the temperature fell to minus twelve degrees." On November 15 a false alarm caused Miles's force to take up battle positions when his scouts spotted Indians in their path, but they turned out to be part of the peaceful village from the reservation near the confluence of the Big Dry and the Missouri at Fort Peck.

When he reached Fort Peck Miles obtained intelligence suggesting that Sitting Bull's band was camped near Black Buttes, about forty miles back near the heart of the Hell Creek region. On November 19 he split his contingent, with part under the command of Captain Simon Snyder moving back up the Big Dry, and Miles moving west up the northern bank of the Missouri before circling back to the southeast in an attempt to surround Sitting Bull. The troops on the Big Dry arrived at Black Buttes on the twenty-seventh to discover that the Hunkpapas' camp was nowhere to be found. Low on supplies and unable to contact Miles, Snyder left for their Tongue River base on December 2, leaving a trail of their dead livestock in their wake. They arrived on the tenth, after marching 330 miles in a futile chase.

Meantime, Miles was experiencing more exasperation along the Missouri north of Hell Creek. After strong currents studded with ice floes nearly capsized the raft and makeshift wagon-boat the troops constructed to ford the Missouri, Miles decided to move farther upstream where he might cross more safely on the iced-over river. But before he could leave he received a report on November 29 that Sitting Bull had actually gone east from Fort Peck, rather than west. While Miles was courting death 110 miles west of Fort Peck, Sitting Bull's band of 170 lodges was safely ensconced 30 miles east of it.

As Greene probably understates, this news was both "disconcerting and frustrating for Miles." Sitting Bull had managed to outflank Miles's trap, raising the dangerous prospect that bands of Sioux led by Sitting Bull and Crazy Horse could reunite south of the Yellowstone River.[30]

To counter that potential development, Miles sent Lt. Frank

Baldwin and three companies back to Fort Peck with orders to try and track down Sitting Bull in the region east of the fort.

Meanwhile, Miles and his remaining troops marched about twenty miles upstream before crossing the Missouri and beginning back east. They then headed south along Crooked Creek, which forms the western edge of the Hell Creek region proper. The rugged breaks held both assets and liabilities. The soldiers struggled to move the wagons on the third and fourth, often having to stop and cut roads before proceeding, but game was abundant, providing a feast of venison. They were passing through some of the richest dinosaur exposures in the world, but nothing could be further from their minds. Once past Squaw Creek, they bore southeast through the heart of Hell Creek and toward the upper drainage of the Big Dry. Feed for the livestock was running low when, on the seventh, they spotted Black Buttes. As temperatures once again plummeted, they strained to traverse the broken ground along the Big Dry and its tributaries, before following the trail of dead livestock left by Snyder's detachment which had earlier returned up that stream course. On the twelfth Miles marched late into the evening, as a snowstorm enveloped the column and the scouts in the vanguard had to fire their rifles so the line of troops and wagons would know where to follow. Food and forage was perilously low; as one soldier recalled, "Our mules grew so thin . . . that you could almost read a newspaper through them." But a relief column arrived with supplies on the thirteenth and by the fourteenth, Miles reached the base at Tongue River after a grueling trek of 558 miles in five weeks.

Meantime, Baldwin and his 113 troops reached Fort Peck and, after resupplying, struck out to locate Sitting Bull, whose one hundred lodges were camped along the north bank of the Missouri to the east. Greene concludes that Sitting Bull and his followers still found the region around the Big Dry to be the "primary attraction" because of the area's abundant supply of bison and other game during the winter.[31] On Baldwin's way there, fresh intelligence indicated that Sitting Bull and his entourage had crossed the Missouri near Bark Creek and camped on the Milk River.

On December 7 Baldwin fought a skirmish against Sitting Bull's force, estimated to be six hundred strong, near Bark Creek, as the Hunkpapa retreated to the south bank of the river. Fearing ambush by superior forces, Baldwin called off the attack and returned to Fort Peck, while Sitting Bull moved farther south toward the Yellowstone. Baldwin later recalled that the night march back to Fort Peck was especially brutal, as a "norther" descended, dropping temperatures to thirty-five below zero; "all along the march every officer was constantly busy watching that some man should not lay down from stupification caused by cold and then die."[32]

After again briefly regrouping, Baldwin resumed the pursuit toward the southeast, sneaking up on Sitting Bull's village of 122 lodges nestled beneath a bluff along Ash Creek near the divide between the Missouri and the Yellowstone. While most of the warriors were out hunting, Baldwin's troops routed the village, burning most of the tepees and capturing sixty ponies and mules, as well as food and buffalo hides. After the victory,

Baldwin headed immediately back toward the base on the Tongue, as incensed but poorly supplied warriors harassed them. Arriving December 21, Baldwin's regiment had covered an unbelievable 716 miles in just over six weeks through the badlands and blizzards of the Hell Creek region.

Ironically, this pivotal battle at Ash Creek took place just a few miles north of the spot where Miles had started his attack at Cedar Creek. Basically, Miles and his troops had chased Sitting Bull more than seven hundred miles after the first engagement, to wind up at essentially the same spot, just below the divide that separated the Yellowstone and the Missouri. Although this seems silly, it probably reflects Sitting Bull's love of the area, especially his knowledge that bison were plentiful near the divide.

Battered and beleaguered, Sitting Bull moved south across the Yellowstone, seeking to join forces with Crazy Horse and gain some relief, but bad weather and deep snow drifts prevented the reunion. In January he once again headed north with about 200 lodges of Hunkpapas, Miniconjous, and Sans Arcs, possibly through the Big Dry drainage, on his way to the Canadian border. As Greene relates, the British government in Canada reported that an entourage of 109 Sioux lodges, which would later expand to 200, entered their territory near Wood Mountain. While crossing the border, Sitting Bull and his associates expressed "their intention to remain there peacefully and permanently."[33]

Miles's dogged pursuit of Sitting Bull, as well as Crazy Horse, certainly turned the tide of the Great Sioux War. Never again would the legendary Hunkpapa threaten either the army or the settlers of the northern plains. As a result of Miles's campaign, Sitting Bull clearly lost considerable influence with his brethren and allies. But Miles never did achieve his ultimate goal. He never captured the elusive warrior on his home turf of Hell Creek. Not until July 1881, after the beloved herds of bison had been decimated, did Sitting Bull lead forty-three families back across the Canadian border to surrender, not to Miles, but at Fort Buford, near the confluence of the Missouri and Yellowstone. All the other tribes had already capitulated, but, unbowed, as Utley records, Sitting Bull proclaimed to the commanding officer:

> I wish it to be remembered that I was the last man of my tribe to surrender my rifle, and this day I have given it to you.[34]

5

THE VANISHING HERDS

Most of us learned the lines as kids, either in our homes with our families or at school during singing sessions with our classmates. The words of the ballad paint a most romantic portrait of life on the western plains:

Oh, give me a home
Where the buffalo roam,
Where the deer and the antelope play . . .

I often wondered, as a child, what it must have looked like. My parents grew up in eastern Kansas, and each summer we would make the pilgrimage from Los Angeles through the West and the Great Plains to visit relatives. We often saw deer and even antelope when our route veered through the northern plains, but rarely buffalo (or bison, as they are more properly called). Only in Yellowstone National Park and a few enclosures in South Dakota did I ever see live ones in the flesh. What happened?

The demise of the bison is fairly well documented. Without doubt it can be traced to overhunting, but the question of who was responsible is a bit more difficult to answer. Several factors seem to have come into play.

Estimates vary a bit as to how many bison once roamed the plains from Texas up into Canada, but most estimates fall in the range of forty to fifty million. It was these plentiful populations that formed the cornerstone of the western societies of Native Americans. For many millennia the massive herds provided most of the basics that the tribes required, including food from their meat as well as clothing and shelter from their hides. In turn, bison were worshiped as quasi-divinities.

In the end, Sitting Bull and his cohorts bemoaned their loss of freedom, because they could no longer follow the migrating herds to sustain their nomadic lifestyle. That was the root of his resentment, deeply felt and profoundly disturbing, toward Miles and the settlers from the East. Although some tribes, such as the Arikara, operated on an agricultural base, the Sioux looked upon these distant relatives with contempt, often referring to them as "Corn Eaters." Sitting Bull's animosity towards his brethren who accepted the government's edict to settle on reservations was similarly disdainful, as shown by his own words:

You are fools to make yourselves slaves to a piece of bacon fat, some hardtack, and a little sugar and coffee.[1]

Nonetheless, as game became more scarce on the plains, even he had to capitulate in order to save his followers from starvation. The slaughter of the bison herds in Sitting Bull's favorite hunting grounds along the Big Dry in the Hell Creek region, along with other areas on the Great Plains, appears to have resulted from behavior shared by both the Native Americans and the settlers from the East.

In the spring of 1805, as Lewis and Clark labored up the Missouri near the eastern border of present-day Montana, the Corps of Discovery was awed by the abundance of game, including the bison herds. Lewis's journal entry for April 22 is typical:

I ascended to the top of the cut bluff this morning, from whence I had a most delightful view of the country, the whole of which except the vally formed by the Missouri is void of timber or underbrush, exposing to the first glance of the spectator immence herds of Buffaloe, Elk, deer, & Antelopes feeding in one common and boundless pasture. . . . walking on shore this evening I met with a buffaloe calf which attached itself to me and continued to follow close at my heels until I embarked and left it. . . . Capt Clark informed me that he saw a large drove of buffaloe pursued by wolves today, that they at length caught a calf which was unable to keep up with the herd. . . . [2]

As the Corps neared the Hell Creek region above the mouth of the Yellowstone on April 29, Lewis continued to record that game is still very abundant we can scarcely cast our eyes in any direction without perceiving deer Elk Buffaloe or Antelopes.[3]

On May 9, on the day they reached the Hell Creek region proper and christened the Big Dry River, Lewis made clear that bison comprised a key component as well as a favorite food of the Corps' diet. He provided a lengthy description of how the French Canadian Charbonneau prepared a kind of sausage called "boudin blanc," using bison meat instead of the traditional pork or veal. Essentially, the muscle underneath the shoulder blade was chopped up along with the kidneys, and after appropriate seasoning with salt and pepper, the concoction was mixed with a bit of flour and stuffed into about six feet of lower intestine. Once the end is tied off,

> it is then baptised in the missouri with two dips and a flirt, and bobbed into the kettle; from whence, after it be well boiled it is taken and fried with bears oil until it becomes brown, when it is ready to esswage the pangs of a keen appetite or such as travelers in the wilderness are seldom at a loss for.[4]

Ah! Even way back then on the frontier French cooking seemed to have a superiority that garnered rave reviews from the critics; for, as Lewis concludes:

this white pudding we all esteem one of the greatest
delacies of the forrest . . . [5]

Lewis always kept an eye out for practical applications of
the resources he encountered, as Jefferson had requested.
Sometimes, in his enthusiasm, he seemed to stretch the point a
bit, such as his suggestion that bison hair would comprise a
good substitute for wool. But faced with such an overwhelming
abundance of bison, what Lewis and Clark could not conceive
was that within a century there would be neither bison herds
nor Native Americans idyllically roaming the Plains.

That both Native Americans and settlers were wasteful of
the natural resource represented by the bison can be documented by their hunting practices. Encountering an assemblage of more than one hundred carcasses, Lewis described one
Native American method on May 29, as the Corps passed the
mouth of the Judith River just east of the Hell Creek region:

The Indians of the Missouri destroy vast herds of buffaloe at one stroke . . . one of the most active and fleet
young men is selected and disguised in a robe of buffaloe skin . . . he places himself at a convenient distance
between a herd . . . and a precipice . . . the other Indians
now surround the herd . . . moving forward toward the
buffaloe; the disguised . . . decoy place[s] himself suffi-
ciently nigh the buffaloe to be noticed by them when
they take to flight and . . . they follow him in full speede

to the precipice, the cattle behind driving those in front over . . . until the whole are precipitated down the precipice forming one common mass of dead an[d] mangled carcases: the decoy in the mean time has taken care to secure himself in some cranny or crevice of the clift which he had previously prepared for that purpose.[6]

He goes on to note that the warriors then carve out as much meat as they want and abandon the rest to the wolves, leaving a most dreadful stench.

A variation of the method described above, which is often referred to as a "buffalo jump," was described later in the Smithsonian's annual report for 1887, which included excerpts from Catlin's *North American Indians*.[7] This approach, dubbed the "surround," was used when a suitable precipice was not available or desired. Tribesmen, mounted on horses and armed with bows and arrows or long lances, basically surrounded a herd without its knowledge, at a mile or more in distance, and gradually closed in from two sides. Once the herd perceived their presence, they stampeded in one direction, then another, but other lines of Indians blocked the bison's escape. As the circle closed, pandemonium broke out, and as the bison panicked, the hunters slaughtered the herd.

In this way this grand hunt soon resolved itself into a desperate battle, and in the space of fifteen minutes resulted in the total destruction of the whole herd. . . .

It is to be noticed that every animal of this entire herd of several hundred was slain on the spot, and there is no room to doubt that at least half (possibly much more) of the meat thus taken was allowed to become a loss. People who are so utterly senseless as to wantonly destroy their own source of food, as the Indians have done, certainly deserve to starve.[8]

The Minitarees, Mandans, Cheyennes, Arapahoes, Sioux, Pawnees, and Omahas were all reported to employ this method of "surround" in their hunting. Relying solely on such accounts, one might be tempted to conclude that Native Americans suffered a similar fate to that of Odysseus' crew, after they slaughtered the sacred oxen of the sun god.

However, the commentary in the annual report seems seriously one-sided. Perhaps the administration was still angry at Custer's defeat a decade before, because the death knell for the bison clearly came from the hands of invaders from the East, not the Native Americans. In an article published by *The River Press*, dated August 24, 1977, the hunting done before the Civil War is documented to have started the downfall. Against Native American resistance,[9] as many as four thousand bison hides were sent down the Missouri each year, even though there was little interest in the European market for buffalo robes. But, about 1840, at the same time that beaver hats went out of fashion, robes of bison fur came into vogue. Commerce in these robes expanded to the point that traders in Fort Benton, Montana, shipped about twenty thousand bison hides a year during the 1840s and 1850s. "This grew to about forty-eight thousand by 1857. Most was simply Indian surplus from hunting, and the big animals were plentiful all around."

But the carnage wrought by buffalo hunters accompanying the crews that built the transcontinental railroad lines would change that. One historian calculated that the signing of the transcontinental railroad act in Civil War days was the death warrant for the buffalo."[10] To feed the railway workers, hundreds of crews with hunters, skinners, and teamsters fanned out among the herds, each crew killing up to five thousand bison a season. Armed with "Sharps long range, heavy caliber rifles," their reign of carnage among the southern herds along the Republican River, the Arkansas River, and the Staked Plains of Texas left only isolated strays by 1879. The next year the northern herd came under attack, spurred on by the army's belief that exterminating the bison would inevitably lead to the downfall of the Native American tribes that relied on them for food. Accordingly, "immense kills were made by five thousand whites shooting and skinning in the triangle of the Musselshell, Missouri, and Yellowstone. One Glendive dealer shipped two hundred and fifty thousand hides in one season." This triangle includes Sitting Bull's favorite winter hunting grounds along the Big Dry at Hell Creek. By the beginning of the 1881–2 season plentiful herds still roamed the region around the Bighorn Mountains and eastern Montana, but they, too, had been nearly eradicated by 1883.

When word of the diminishing herds began to reach the East Coast of the United States, a thirty-one-year-old museum

William T. Hornaday, chief taxidermist for the Smithsonian Institution, working on the mount of a large cat. Hornaday's techniques for mounting animals in lifelike postures and natural environmental surroundings revolutionized the way that natural history museums exhibit their zoological specimens. (SMITHSONIAN INSTITUTION ARCHIVES, RECORD UNIT 95/BOX 13/FOLDER 38/IMAGE #3687)

employee with a deep love of nature decided to raise the alarm. His name was William T. Hornaday, and he had already served as the chief taxidermist at the Smithsonian Institution's Museum of Natural History for four years, and was widely regarded as the nation's foremost expert in his field. As he tells the story in *A Wild Animal Roundup*:

In March, 1886, the writer received a severe shock, as if by a blow on the head from a well-directed mallet. He awoke, dazed and stunned, to a sudden realization . . . that the buffalo-hide hunters of the United States had practically finished their work. The bison millions were not only "going," but gone! . . .

The case was so serious that the writer prepared and handed to Professor Spencer F. Baird and Doctor G. Brown Goode a formal letter setting forth the gruesome facts. . . . However, a belief was expressed that there were, even then, somewhere in the West, some unkilled bands of bison from which specimens might be taken before the last of them were swept away.

Professor Baird at once sent for me, and said:

"The situation as you describe it is most serious. I dislike to be the means of killing any of those last bison, but since it is now utterly impossible to prevent their destruction we simply must take a large series of specimens, both for our own museum, and for other museums. . . . You must go west as soon as possible. . . ."[11]

By now I suspect you can guess where Hornaday headed. Undoubtedly, Sitting Bull would have done the same thing, but even though Hornaday was an avid hunter and outdoorsman, there is a genuine note of regret in his words in assuming this role of executioner. As Hornaday laments:

To all of us the idea of killing a score or more of the last survivors of the bison millions was exceedingly unpleasant; but we believed that refraining from collecting the specimens we imperatively needed would not prolong the existence of the bison species by a single day. . . .

. . . Then, in 1886, there were in all the world fewer than 800 bison alive. . . . [12]

Hornaday goes on to note that well over half of those bison were not plains bison, but rather forest-dwelling bison that lived in the northern reaches of the Canadian wilderness near Great Slave Lake.

In May 1886 Hornaday embarked on a scouting trip to evaluate where his best chance of securing the bison might be. Arriving in Miles City, he was most concerned by the locals' reaction to his quest. He was

greeted at every step by the cheerful assurance, "The buffalo are all gone; and you can't get any anywhere."[13]

Nonetheless, Hornaday persevered, mainly bouyed by the advice from Doctor J. C. Merrill, who contended that about seventy-five head still roamed the remote prairies and badlands of the Big Dry region. Eventually meeting a rancher from that area, Hornaday heard confirmation of the bisons' presence, based on rare sightings earlier in the month. Hornaday assembled a small crew and set out for the area just south of Hell Creek. Ironically, not only was Hornaday headed for Sitting Bull's favorite hunting spot, the wagon and many of the supplies for this trip had been arranged through the army at Fort Keogh. This was the same military base that Miles had founded at the mouth of the Tongue River during his pursuit of the legendary Hunkpapa. But there were sobering signs of the slaughter along the way up the Sunday Creek trail toward Hell Creek:

> From the Red Buttes onward you could see where the millions had gone. This was once a famous buffalo range, and now the bleaching skeletons lay scattered thickly all along the trail. . . .
>
> Now and then you came to a place where the hunter got a "stand" . . . Here you could count seventeen skeletons on a little more than an acre, and nearby were fourteen more that evidently fell at the same time. . . .
>
> Beyond the Red Buttes, we were seldom out of sight of bleaching skeletons, and often forty or fifty were in sight at one time. . . . [14]

Soon after arriving in the Hell Creek region the somber atmosphere was lightened somewhat by a remarkable event. Hornaday and his crew captured a tawny bison calf along Sand Creek, just south of the Big Dry.

> Instead of being dusty brown, like most buffaloes over a year old, he was a perfect blonde. His thick, wavy coat was of a uniform bright sandy color, and 'Sandy' he became from that moment. [15]

The orphaned calf was taken to a nearby ranch where it was grudgingly nursed back to health by a cow while Hornaday's crew continued their reconnaissance. Eventually they shot two bison bulls, but realized that the hides were not in peak condition during that part of the season. So, in June, with Sandy in tow, Hornaday headed back to the Smithsonian in order to plan a full-scale return expedition the next fall. Initially, Sandy did well on a farm near Washington, and when tethered on the lawn outside the museum, he briefly became quite a celebrity with museum visitors:

> With an abundant supply of good food, over two gallons of good milk per day, it grew rapidly, and soon became quite fat. . . . but as has been the case with many other distinguished foreigners, life in Washington proved too rich for his blood.
>
> About the middle of July he ate a great quantity of damp clover, and before anyone found it out he was dead.[16]

By September 24 Hornaday was back in Miles City, along with a science student who would serve as his assistant. In addition, three cowboys joined the crew, as did an army cook from Fort Keogh, which again supplied the team with a wagon, a Sibley tent, and other camping gear. The crew spent the next two months hunting north of the divide between the Yellowstone and the Missouri in an area just south of the Big Dry, about forty miles long from east to west, and about twenty-five miles wide from north to south. Their goal was to collect about twenty bison for the museum collections, but it would be anything but easy. Gone were the days of Lewis and Clark, in which the bison were not suspicious of humans. Connell provides one scenario of the bison's gullibility during earlier days:

> Durable and obstinate he might be, but simple to predict—the cows especially. Father Pierre Jean De Smet watched an Assiniboin approach a herd, conceal himself, and imitate the bleat of a calf, at which all the females hurried toward him. He shot one. The others ran away. The Assiniboin reloaded his rifle and resumed crying. The females stopped. They looked around. As though enchanted they hurried again toward this noise. He shot another, which supplied all the meat he wanted, but he assured Father De Smet that he could have gone right on crying like a baby and killing them.[17]

Yet even bison eventually respond to the selection pressures inherent in Darwinian evolution. Hornaday and his fellow hunters had a difficult time even spotting any bison in the rugged terrain south of the Big Dry. For two weeks:

> We had laboriously hunted the country on both sides of Sand Creek, but saw no signs of buffalo until the 13th of October . . . one of our cowboys . . . came upon a bunch of seven buffalo lying in the head of a deep ravine, but although he fired on them several times and chased them two or three miles, they all got away. . . . It

was not long before other discoveries confirmed our surmise that the buffalo we were after were in the habit of hiding in the heads of those great ravines whenever they were disturbed on their favorite feeding grounds . . . [18]

The dilemma that Hornaday felt about being an agent of extinction is clearly acute. On the one hand he bemoaned the slaughter of the previous decades, as documented previously, but on the other hand he was clearly captivated by the spirit of the hunt, which turned into an outright contest:

Jim McNaney, a splendid shot and a genuine buffalo-hunter, with a record of about three thousand three hundred head, slain for their hides in three years, had (up to that point on the trip) killed five of the ten head, while L. S. Russell was credited with three. Boyd and I were behind in the race, and aside from our desire to get buffalo by all possible means, each man was ambitious to keep up his individual score.[19]

In addition to his scientific responsibility to collect specimens for the Smithsonian's collections, Hornaday seemed to justify the hunt as a worthy endeavor because of its present difficulty:

In days gone by, hunting buffalo was tame work, owing to the great abundance of the animals and their stupidity.

There was no more glory in killing an old bull than in wringing a rooster's neck, for familiarity had bred contempt. But with the approach of extermination, and 'the struggle of the species to harmonize with its environment' (by the kind permission of the evolutionist) conditions have changed, and in 1886 the chase of the buffalo was sport of the very toploftiest kind.[20]

For Hornaday, persistence finally paid off. Toward the end of the trip the crew moved its camp to the head of Big Porcupine Creek in search of better water, which had been difficult to find, since none of the creeks in the area were running at the time. This spot boasted

a deep pool of delicious water, without alkali and inexhaustible in quantity.... Two hundred yards from the pool...we pitched our new Sibley tent close to the southern face of a perpendicular bluff that formed a semicircular wall around us, and like a sheltering wing protected us at all times from the cold winds that swept across the bleak level of the High Divide.... In the head of a deep gully close by ... we made a very comfortable 'dug-out' to serve Mack as a kitchen....

From every direction save one there came down great precipitous gullies, and from our tent we looked due south down the rugged little canyon for two miles or so to where the view was abruptly cut off by a lofty isolated butte.[21]

Although the butte was dubbed Smithsonian Butte by the crew, it is now known as McGinnis Butte. Recently identified as being situated along the bed of McGinnis Creek, the site of the camp has been acknowledged as a national historic site, and it was from this camp that Hornaday embarked on his unforgettable foray after the specimen that would make him famous. Heading south toward "Buffalo Buttes," Hornaday and McNaney spotted the ultimate prize:

We had crossed two or three ridges on ... the farther slope of the Divide when ... we suddenly espied ... the light brown humps of *three buffs.*

One of the buffs was an *immense* old bull. . . . I fired at the bull ... and then they made off in good earnest. We poured it to them as they ran up a little hill. . . .

The old bull ... ran over the hill and disappeared, and then we rushed for our horses. . . .

I ... gave him a shot fair in the shoulder. Down went the great beast, head foremost. . . .

... he staggered to his feet, in spite of his broken leg and galloped off over the hill. . . . After a short run we again overhauled our prize, on the side of a hill, near the crest of which he once more halted and stood at bay. Thirty yards away from him I pulled up, and gazed upon him with genuine astonishment. . . . He was a perfect monster in size, and just as superbly handsome as he was big. In his majestic presence ... I thought to myself:

"Until this moment, I have never had an adequate conception of the great American bison."

. . . I studied his lines with absorbing interest, and took one mental photograph of him after another. . . . Several times his head sank very low, and he viciously pawed the wet snow with his wounded foreleg. But these intervals of anger would pass away, his eyes would lose some of their fire, and he would content himself with simply regarding me.

With the greatest reluctance I ever felt about taking the life of an animal, I shot the great beast through the lungs, and he fell down and died.[22]

Hornaday admitted that this was the most impressive four-legged animal he had ever seen, including lions, tigers, and elephants. Estimated to be nine years old and, thus, probably born the same year that Miles chased Sitting Bull through the region, it stood five-feet-eight inches at the shoulder, extended nine-feet-two inches in length and weighed 1,700 pounds. It had clearly been hunted before, as evidenced by four bullets in its body, in addition to the ones fired by Hornaday that had failed to kill it. From the extensive running to avoid hunters, its body was exclusively sculpted in muscle without any fat. His pelt was perfect, with a mass of blackish-brown, curly tufts adorning the massive head, and six-inch-long straw-colored hair covering his hump and shoulders. Hornaday drew several sketches of the fallen patriarch, for he had ambitious plans to immortalize him. Those plans would take shape over the next year. With the goal of twenty bison for the collections realized, Hornaday left the Hell Creek region in mid-December and set out across country to the Smithsonian. By the next year, when a field crew from the American Museum of Natural History (where I worked a century later) returned to this territory to collect specimens for their collection, none could be found.

Upon his arrival in Washington Hornaday launched into an innovative exhibition project that would forever change the way natural history museums portray their collections. Up to that point exhibits of animals generally consisted of single specimens mounted in a static pose on a horizontal wooden base. Few expressions or dynamics of the animal were even attempted. But in homage to the bison that had so troubled and affected him, Hornaday would change that paradigm.

Using six of the specimens collected near Hell Creek, Hornaday conceptualized and executed the first exhibit of a genre now referred to as a life or habitat group. The members of Hornaday's cast included the trophy bull, Sandy, and four other specimens representing different ages and genders. His goal was not only to portray the diversity within the species, but also to show them as he had seen them in the wild. He designed a display in which all the animals would be featured in natural positions around a water hole, complete with the interactions that would commonly follow. In order to make the scene more realistic, Hornaday had shipped barrels of soil and even samples of turf from the prairie in the Hell Creek region. He took great pains to utilize every trick and technique he had

Exhibit of William T. Hornaday's habitat group featuring six of the bison specimens he collected in the region south of Hell Creek in 1886. The exhibit opened to the public at the Smithsonian Institution in 1888. (SMITHSONIAN INSTITUTION ARCHIVES, RECORD UNIT 95/BOX 43/FOLDER 1/IMAGE #4475)

learned, both in the field and in his taxidermy studio, including using the feet of the specimens to create tracks in the mud and dirt near the pool. The result created a sensation, both within the media and the public. As described by Harry P. Goodwin in the Washington *Star* on March 10, 1888:

A little bit of Montana—a small square patch from the wildest part of the wild West—has been transferred to the National Museum. It is so little that Montana will never miss it, but enough to enable one who has the faintest glimmer of imagination to see it all for himself—the hummocky prairie, the buffalo grass, the sagebrush, and the buffalo. It is as though a little group of buffalo that have come to drink at a pool had suddenly been struck motionless by some magic spell, each in a natural attitude. . . . The finishing touches were put on today, and the screens will be removed Monday, exposing to view what is regarded as a triumph of the taxidermist's art.[23]

According to Douglas Coffman, writing in *The Vanishing West*, when a Sioux dignitary visited the exhibit he was so taken in by it that he thought the bison must wander around inside the case at night, "because their tracks were plainly visible."[24]

Hornaday's genius for resurrecting the skin of dead animals back into living beasts not only entertained but educated the public about the animals' habits and the environment in which they lived. This quantum leap in exhibition technique still reverberates through the museum world. From Hornaday's life group, it was a small step to add a background mural and create the exquisite dioramas that have become the cornerstone of modern natural history museums. Even when I directed the renovation of the fossil halls at the American Museum of Natural History in New York during the late 1980s and early 1990s, we drew inspiration from Hornaday's precedents in remounting our dinosaur skeletons in dynamic lifelike postures.

So powerful was the scene that Hornaday's majestic bull became not only an icon for the museum community but also a symbol of the United States government. The bull still forms the focal point on the Great Seal of the Department of the Interior, which manages all our national parks. It also was portrayed on a ten-dollar bill that was printed in 1901 and may have been used in part as a model for the buffalo-head nickel. Three different postage stamps have also borne its image.

But Hornaday wasn't finished working with bison. He became the designer and founding director of the New York Zoological Park, now know as the Bronx Zoo, in 1896. His primary policy goal there was to promote wildlife conservation and protect endangered species through breeding programs at the zoo, including the bison.

Even before Hornaday's Montana hunts, inadvertent steps were being taken that would lead to the salvation of the bison. As told in *The Vanishing West*, a man, named Samuel Walking Coyote, living on the northern plains of Montana, adopted eight bison orphans that wondered into his camp in 1873. By 1884 his herd grew to number thirteen, and he sold the herd to two other men. By 1895 the herd's population had multiplied into the hundreds. Eventually, fifteen were sold to the U.S. government to stock Yellowstone National Park, and seven hundred were sold to Canada.

Hornaday also bought some bison from the herd that Samuel Walking Coyote had nurtured. As Coffman writes in *The Vanishing West*, it was during his tenure in the Bronx that Hornaday played a pivotal role in organizing the first bison preserve in Oklahoma, called the Wichita, which was stocked with animals from Hornaday's herd at the zoo. Later, in 1905, he became the founding president of the American Bison Society, a role in which he fostered the establishment of several national bison ranges. The first was sanctioned at Moiese,

Montana, in 1908. The bison released in the Moiese Range also descended from Samuel Walking Coyote's orphans.

The portrait we have of Hornaday is certainly not monochromatic, but intensely complex. Clearly, Hornaday felt conflicted about the role he played in the bison's natural history. It is not really fair to judge him from all the perspectives of environmental awareness that we hold today. He was a man of his time, yet more enlightened in many ways, in terms of both artistry and conservation. How should we judge a man who hunted the last remnants of the great North American bison herds to the brink of extinction in the name of science, but then worked tirelessly to fend off the ultimate demise of the species? Coffman seems to sum it up best:

> There was a definite direction in his life's work which is manifest in his writings. Although he was a zealous sportsman, Hornaday clearly tempered his love for adventure with a respectful reverence for the world of nature. He rued the wasteful slaughter of wild animals, and worked vigorously to prevent their decline.[25]

The National Register Nomination for Hornaday's camp goes on to conclude:

> Through his many publications and personal efforts, he became the driving force behind key national and international laws and treaties which provided protection for wildlife. . . . There is no doubt that Hornaday was the

father of wildlife preservation in America and the world, for that matter.[26]

Hornaday's bison exhibit stood as a highlight in the Smithsonian's mammal hall for seventy years, until 1955, when new specimens were installed during a renovation. As Coffman relates in *The Vanishing West*, when Smithsonian workers disassembled Hornaday's exhibit in 1957 they discovered a metal box in the base of the display case. The box contained a copy of Hornaday's article published in *The Cosmopolitan* in 1887, which he titled *The Passing of the Buffalo*. His handwritten note embellished the cover page:

To My Illustrious Successor,

Dear Sir: Enclosed please find a brief and truthful account of the capture of the specimens which composed this group. The old bull, the young cow and the yearling calf were killed by yours truly. When I am dust and ashes I beg you to protect these specimens from deterioration and destruction. Of course they are crude productions in comparison with what you produce, but you must remember that at this time (A.D. 1888, March 7) the American School of Taxidermy has only just been recognized. Therefore, give the devil his due, and revile not.

W. T. Hornaday
Chief Taxidermist, U.S. National Museum[27]

The six specimens of Hornaday's group were returned by the Smithsonian to Montana and eventually dispersed. But through a concerted effort in the late 1980s and early 1990s by Douglas Coffman and the curators at the Museum of the Upper Missouri, the specimens have been relocated and reconserved for display to the public in Fort Benton. In the end, Hornaday not only saved the species he had come to respect so much from extinction, he had immortalized many of those he had killed through his artistic genius.

6

THE RESURRECTION OF
TYRANNOSAURUS

In addition to the enduring love and respect that Hornaday exhibited throughout the rest of his life for bison, he harbored the same emotions for the landscape that had fostered and supported them. Writing in *A Wild Animal Roundup*, Hornaday romantically relates tales of his return to the Hell Creek region, when he briefly reunited with Jim McNaney and the widely renowned western photographer, L. A. Huffman. Their quest on this trip was quite different from that which drove the bison expeditions of 1886. On this foray, they intended to hunt and document the lifestyle of the blacktail deer, deep in the badlands of Hell Creek, rather than chase bison across the higher prairies and buttes slightly south. Although Hornaday states that this trip took place in 1908, I suspect that some of the events took place in October 1901, based on reasons that will become clear later.

In any event, the crew gathered once again in Miles City and headed north up Sunday Creek for the Big Dry. The area around the Big Dry had changed considerably in the fifteen or twenty years since the Smithsonian's bison hunt. Distinct symbols of "civilization" had become obvious to his trained eye. As the horsemen and wagon passed over the Divide, Hornaday noted a disturbing change in the landscape:

> In days gone by that was one of the finest buffalo ranges in all the West. After the buffalo days, this side of 1884, it was a fine cattle range; but the awful sheep herds have gone over it, like swarms of hungry locusts, and now the earth looks scalped, and bald, and lifeless.[1]

Fording the Big Dry, which Hornaday, in apparent surprise, noted was "a river of real water," the party confronted another novelty, Jordan. It wasn't much of a town, but Hornaday notes the presence of a store and a post office. Both those establishments owed their existence to an intrepid English immigrant.

Arthur Jordan was a restless soul. Born three years before the battle of the Little Bighorn, he was fourteen when he left his family near Wolverhampton in Britain for a life of adventurous adolescence, which he recounts in his autobiography, *Jordan*. After serving as an underaged roustabout on steamships along the coasts of Ireland and the British Isles, he worked his way across the Atlantic on the steamer *Freeman* at the age of about fifteen. By 1889, three years after the Smithsonian bison expedition, he had drifted west to Deadwood in the Black Hills

Wedding portrait of Arthur J. Jordan and Hattie Elizabeth Jordan taken in 1894. (COURTESY OF THE GARFIELD COUNTY MUSEUM, JORDAN, MONTANA)

area that the Lakota Sioux revered so highly. Jordan funded his travels with odd jobs along the way. He took a job working along the stage line between Deadwood and Miles City, the town that quickly grew up around the military base that Miles had started and used to support his campaign against Sitting Bull. Dubbed "The English Kid" by the drivers along the stage route, Jordan spent a considerable amount of time filling in for them during their frequent drunken binges as they traveled through the Powder River drainage. He even befriended the Cheyenne chief Two Moons after the massacre at Wounded Knee, when Jordan helped the army escort the tribe to the Fort Keogh Military Reservation near Miles City.

Throughout the late 1880s and into the '90s, Jordan ran a few horses and worked as a hand in the cattle operations in the Powder River region. Violence was common as new settlers arrived in the region to set up livestock ranches and to fence off their claims. These restrictions across the open ranges on which the large cattle operations depended climaxed in the Powder River Invasion near Buffalo, Wyoming, in 1892, in which the open-range stockmen, who owned and managed the larger herds, attempted to murder or drive the small-ranch settlers from the region. A large contingent of armed settlers responded in kind, and the army was ordered to step in to quell the disturbance.

In the fall of 1891 Jordan teamed up with a former bison hunter named Charley Hart to explore the same region where Hornaday had hunted five years before. The bison were all but gone; however,

> Antelope could be seen in any direction we gazed, as well as coyotes, lynx-cats, and the big lubberly gray wolves plus a few buffalo that had escaped the ruthless slaughter of a few years before.[2]

Camping just north of the Big Dry, Jordan and Hart discovered one of the difficulties inherent at Hell Creek. After setting camp, they turned their horses loose to graze, but the mounts quickly took flight for no apparent reason. The camp was nestled at the base of the previously mentioned hill named Smoky Butte, whose nooks and crannies were infested with rattlesnakes preparing their winter haunts. The horses had sensed the presence of the snakes through their rattling and scent before most sensibly evacuating the area. Climbing to the top of the butte to search for the missing horses, Jordan waded through dozens of the vipers, reeking havoc as he went. Rattlesnakes still inhabit the region, as previously mentioned: We encounter a few every year as we prospect the buttes and coulees for fossils. They seldom bother you unless you get close enough to threaten them, at which time their rattling sends a primal chill down your spine.

But Jordan was undaunted by the snakes, and after his reconnaissance with Hart, clearly fell in love with the lushly grassed prairies and forested breaks surrounding Hell Creek. Upon returning to the Box Elder region of southeastern Montana, where Hart set up his own spread, Jordan met Hart's niece when she came for a visit in the spring of 1894. He was smitten at first sight, but Ms. Hart was considerably more coy. Nonetheless, Arthur barged ahead, and when a neighbor counseled that he take good care of the girl on their way home after a visit, he replied that that was precisely his intention, while portentously adding,[3] "I shall take good care of her as long as I live."

Later on the ride home the girl protested Jordan's presumption by "wrathfully" exclaiming, "You've got some nerve!" After a startled Jordan asked why she was angry, Ms. Hart queried, "Why did you tell Mrs. Wickham that you were going to take care of me as long as you lived? You had no right to say that." Upon revealing that his profound regard for her had led to his statement, the unsatisfied girl renewed her attack:

"You're impossible," pouted the girl. . . . "When I first met you in the pines, before I had known you twenty minutes, you even tried to kiss me. Do I look like a girl that a stranger could take liberties with?"

Apparently overcome by a measure of contrition mixed with strategic diplomacy, Jordan admitted that he was "ashamed of himself." But she continued to press her defense:

"Do you try to kiss every girl you meet?"
"You are the only girl I have met so far."

An occasionally stormy Western romance ensued, but the couple was married by August of that same year. The next spring Jordan worked with a horse outfit in northeastern Wyoming before returning to his new wife and father-in-law in the summer. Once again growing restless, Jordan pined for more open spaces as the country around him grew settled. In the spring of 1896 he popped a follow-up question:

"Girlie . . . I know a country that is a little more than two hundred miles to the north, where you and I should go and build our home. We would have the whole country to ourselves, grass everywhere knee-high to a horse."[4]

After continuing to describe the remoteness of that countryside, replete with its rugged breaks, spacious prairies, and

abundant wildlife, Jordan declared that it represented the ideal home for the two of them.

"Shall we move into that country, Girlie?"

"Just as you say," she answered. "I shall go with you, no matter where you go."

So the newlyweds packed for Hell Creek and the range that both Sitting Bull and Hornaday had trod with such reverence. In June of 1896 Jordan and his wife headed north to Miles City to gather more supplies before fording the Yellowstone and traveling upriver to Forsyth. From there they turned north up the valley of the Big Porcupine River, camping for awhile near the headwaters of both that stream and Sand Creek. They were essentially in the same country near McGinnis Butte, where Hornaday had camped during the last phase of his bison hunt in 1886. One decade later, Jordan reports that there were still three small bands of bison inhabiting the region, but by 1897 all had been killed by a small band of hunters that rode through the region.

While based near the head of Sand Creek, Jordan recounts,

I drove about twenty miles down Sand Creek, then over to the Big Dry. I had my wife with me, and we selected the place where we would build our future home. There were no wagon roads, so we drove over the uneven ground, through the sage brush, and across the cuts which I shoveled to make the crossings easier.[5]

Clearly, road construction had not progressed much in the twenty years since Miles had scurried around the region during his fruitless chase.

During the end of October and the beginning of November, Jordan, his wife, and their three-week-old daughter moved their camp to the bank of the Big Dry where the town of Jordan now stands.[6] After setting up camp Jordan's wife emerged from the tent to discover a wolf on the prowl just fifty yards away. Attracted by the aroma of newly butchered venison, the wolf soon met its end, as Jordan sprung into action by shooting it and stripping its pelt.

The next several days were devoted to chopping down trees for their new cabin. Soon the walls began to rise and the split roofing was fastened in place. When a blizzard ominously threatened, Jordan quickly fashioned a sod chimney and moved the family inside where they improvised their furniture from boxes and other makeshift materials.

My diary also records that the first night in the cabin, so roughly put together, rang with song and laughter well into the night, while outside the blizzard raged.[7]

Typical trials and tribulations followed during the wretched winter and subsequent spring, as the small family struggled to fashion their future on the newly settled frontier of Hell Creek. Jordan brought his horses up from the Box Elder region and planted a garden for vegetables as well as fields of musk and

watermelons. On occasional excursions Jordan and his family explored the rugged breaks and bottomlands of the Missouri adjacent to Hell Creek, discovering the myriad of game animals that served as their only neighbors.

> We saw many mountain sheep and blacktail deer in the timber and brush along the river bottomland, and fantail deer in the . . . meadows and in the sagebrush and greasewood back from the cottonwoods. Surely a game paradise yet not a real sportsman's paradise. Nature's wild things were too tame, for they had not yet learned the ways of man.[8]

Over the next few years, however, Jordan's wonderland of splendid isolation shattered as several new groups of settlers descended on Hell Creek:

> In the summer of 1900, the sheepmen and ranchers met at my ranch to sign a petition, asking the postal authorities to establish a post-office at my ranch, the post office to take my name, and with me as postmaster.[9]

At first it was not a terribly sophisticated operation. As told in *Trailing Through Time*, a 1999 publication of the Garfield County Historical Society, when the mail arrived at Jordan's ranch via Miles City over the Sheep Mountains, ". . . it was dumped in a corner and you helped yourself to your mail." In the spring of 1901 Jordan took the next quantum leap in community development when he cut down more cottonwood trees for building materials to erect the town's first store. It had dawned on Jordan that, with the population growing, "he could sell supplies in this area if he brought in extra when he purchased his own." But this new lifestyle did not conform to the expansive boundaries of Jordan's restless soul:

> As I became very busy hauling out supplies for the store, I realized I had changed my mode of living. No more was I a free rider of the range, for now I was a postmaster and storekeeper, and it did not set very well on my shoulders.[10]

With a post office and store established, it apparently seemed that only one essential institution was missing in the newly established town, and the locals used a trip that Jordan and his family took to visit relatives to remedy the situation. When he returned:

> I was somewhat surprised to find that three saloon men had turned the post office into a saloon. I had a number of extra house logs piled up nearby, so I told the men to take the logs and build a saloon building.[11]

With all able-bodied men enthusiastically chipping in, the town's first saloon quickly took shape, but as Jordan himself laments, this establishment "was the source of much trouble in the new town."

It was upon this kernel of civilization that Hornaday stumbled in the fall of 1901. Jordan confirms[12] that Hornaday's hunting party included one of the cowboys who had participated in the Smithsonian's bison hunt, Jim McNaney, as well as the famous photographer of the American West, Laton A. Huffman. The successful trip resulted in the procurement of several "large game heads" and "a number of very fine pictures." But that was not all that Hornaday and Huffman had found during their sojurn at Hell Creek, hosted by a crusty, but in the end personable settler named Max Sieber, a former bison hunter and Texas cowboy who had staked his claim near the edge of the breaks. In addition to hunting, as Hornaday relates:

> Over in the easterly badlands . . . Max found three chunks of fossil bone which when fitted together formed a hornlike mass nearly a foot long. . . . Then Max Sieber took me to a spot near by where he had found the badly weathered remains of what once had been a fossil skull, as large as the skull of a half-grown elephant. It lay quite free, upon the bare earth, in a place that looked very much like the crater of a volcano, it was so blasted and lifeless, and cinderlike. The skull was so badly weathered that nothing could be made of it, but near it lay several fragments of ribs in a fair state of preservation.[13]

So, by 1901 terrors from beyond the bounds of human history—dinosaurs—began to reemerge from the badlands at Hell Creek. They were soon resurrected by intrepid paleontologists who performed the Herculean tasks necessary to extricate their skeletons from the sandstones and siltstones long-buried beneath the surface.

The best dinosaur collector on the early expeditions to Montana was a dapper but determined paleontologist named Barnum Brown, who worked for the American Museum of Natural History in New York, the same institution for which I worked ninety years later. Brown grew up on a farm in eastern Kansas, in the same area where my parents grew up and where we returned every summer to visit our relatives. As a boy of sixteen, Brown accompanied his father on a roundabout wagon trip to Montana in search of a new ranch site—an odyssey of more than three thousand miles.

His love for the landscape and a visit from Hornaday prompted his return to Montana at the age of twenty-nine to search for fossils. Well aware of the interest in dinosaurs at the museum, Hornaday showed his photographs to Brown upon his return to New York. Brown identified some of the remains as belonging to the horned dinosaur named *Triceratops*. In letters that followed their meeting in May 1902 Hornaday did what he could to pinpoint the locations of the discoveries for Brown. One of those letters described the locality for bones found near the small town of Forsyth:

I am just in receipt of a letter from . . . Mr. L. A. Huffman . . . giving the location of the dinosaur found by Mr. Harrison. I send you Mr. Harrison's map of the location,

The young Barnum Brown (left) and his boss Henry Fairfield Osborn in Aurora, Wyoming, during Brown's first field trip for the American Museum of Natural History in 1897. (AMERICAN MUSEUM OF NATURAL HISTORY LIBRARY, IMAGE #17808)

from which I am sure you will have no difficulty in locating it on one of the land-office maps of Montana. It will not, however, be quite so easy to find it on the ground! But you will manage that.[14]

Undaunted, Brown set out the next month by train to Miles City to investigate. Montana was, in many ways, still a vestige of the Wild West. It had achieved statehood in 1889, less than thirteen years before Brown set out from New York. According to the census in 1900, Montana sported a population of only 243,000—an especially meager figure in contrast to the burgeoning population of sheep, which totaled six million, and made the state first in the nation for wool production. A dozen eggs cost fourteen cents, while sugar was four cents a pound, and although the population on the plains of Montana was increasing disproportionately in relation to other areas of the state, Brown was heading toward a particularly remote and untamed terrain.

His ultimate destination was the Hell Creek region, but not the prairies then littered with bison skeletons. Instead, he would seek out the rugged ravines etched into the Great Plains by the mighty Missouri River and its veinlike labyrinth of tributaries. There the buttes and ridges formed by ancient sediments laid down in long-lost streams and seas extended for over one hundred miles along the southern banks of the river that Lewis and Clark first mapped a century before.

Brown was quite comfortable at socializing in any setting, and he quickly got wind from the locals that other dinosaur skeletons abounded between Miles City and the small town of Jordan to the north. On June 17, from Miles City, Montana, he wrote back to his boss at the museum, Henry Fairfield Osborn:

> Yesterday, I heard of another specimen having been found within five miles of Forsyth. I rode out to the place finding the location and the debris, the skeleton having been destroyed by souvenir hunters. It was a Claosaurus [an old name for a kind of duckbill] . . . The vertebrae had been broken off and packed away while as near as I can learn the skull [and neck] vertebrae were broken off . . . and sold to the Smithsonian Institute about four years ago. It seems a Baptist Sunday School teacher had stolen the skull and sold it.[15]

Clearly, Brown already faced competition in this fossil-rich frontier. Since 1877, when the prominent paleontologists Edward D. Cope of the Academy of Natural Sciences in Philadelphia and Othniel C. Marsh of Yale initiated their feud over newly discovered dinosaur skeletons from the American West, institutions vied bitterly for the best collecting localities and specimens. It was commonplace for crews to gather intelligence from the locals about the activities of their competitors, and more than one territorial row arose between crews from different institutions. But none seemed to be around at the moment, and by June 19 he located the "dinosaur" that Harrison had found:

> It turns out to be a mosasaur . . . all badly weathered and broken up . . . not worth bringing in. From the invertebrates associated with this specimen the horizon seems to be Fort Pierre.[16]

Unfortunately, one of Hornaday's specimens that triggered Brown's trip turned out to be a marine reptile from the brown marine mudstone, called the Bearpaw Shale, that underlay the dinosaur-bearing beds, rather than the dinosaur Brown sought. But that was a small setback; for this was not the area in which Brown was primarily interested. He immediately set about to equip himself for a summer-long assault on Hell Creek. Brown mused over the best logistical approach, to impress upon Osborn that he would not be profligate in his preparations.

> Whether to buy or rent an outfit [wagon]. The best figure I have been able to obtain is six dollars per day for rent of an outfit while it is equally expensive to buy, but I rather think I shall buy three horses and rest of equipment at Miles City . . . I shall try to rent an outfit with understanding that rent goes on payment of purchase if I wish to buy after two weeks.[17]

By July 7 Brown and his crew had reached the town of Jordan, now bustling with Jordan's post office and store, as well as a saloon, restaurant, and hotel. But the myriad of tasks involved in managing his first full-scale expedition was clearly taxing Brown's organizational skills, as he apologetically confessed to Osborn:

I wrote you from Miles but discovered the letter still in my pocket yesterday so I will try a new one.[18]

Brown's primary colleague on this trip was one of Osborn's doctoral students, Richard Swann Lull. Tall and giftedly athletic, Lull was six years Brown's senior. He had worked in 1899 at the American Museum of Natural History's Bone Cabin Quarry in Wyoming, as part of Osborn's field crew, after Brown had left for a two-year expedition to Patagonia. Despite Lull's seniority in age and academic qualifications, Brown had more field experience and clearly headed up the Montana crew. Based on their correspondence, they respected one another greatly and became close colleagues. As the result of this trip, Lull would become especially intrigued with horned dinosaurs, eventually publishing several scientific papers and a long monograph about the group.

Jordan clearly met Brown and Lull when they came through the newly founded town. As Jordan recalls in his autobiography, Brown and Lull showed up with their gear in early May 1902—a month or two earlier than Brown actually arrived—

and quickly set out for the breaks. Jordan rode out on several occasions to see their activities for himself:

> I had seen many petrified bones in my ranging through the breaks, but these scientists knew where to dig to obtain perfect specimens of the huge prehistoric monsters. It was very interesting for me to view the specimens.

Jordan adopted a rather protective attitude about the paleontologists. He knew that they could take care of themselves quite well out in the breaks, especially with the help of Max Sieber, who they hoped would play host; but he had good reason to worry about them when they came into town for supplies:

> Those times in Jordan were wild, and every shady character that could not stand the spotlight of civilization drifted in and around the new town, always ready to have fun or start trouble, and a few of them desired to pick on the scientists and their men.

All in all, Jordan was pretty successful in insulating Brown and Lull from the town's hooligans. But one day, when Lull rode in to pick up the mail, Jordan could sense that "trouble was brewing." He warned Lull to steal out of town as inconspicuously as possible so as not to draw the attention of any rowdy inhabitants. Lull gratefully followed Jordan's advice and was almost out of sight before the antagonists realized he had flown the coop.

> They ran for their rifles and began to throw lead all about him. The fellow made his team do their best in going over the hill and away from those drunken, demented morons. Thereafter, the fellow gave the town of Jordan a wide berth.[19]

Amusingly, neither Brown nor Lull ever mentions this incident in their field notes, at least to the best of my knowledge.

By July 12 Brown and Lull had safely reached the ranch north of Jordan, formerly owned by Hornaday's old friend Max Sieber. The road to the ranch dropped down off the parched plains into the sizzling, summer maze of the breaks. Both Brown and Lull were anxious to initiate their search for fossils. In 1902, just as today, collecting in previously unexplored terrain involved two distinct operations. The first is called "prospecting," which requires hiking along the ridges and ravines with your eyes glued to the ground in search of bone fragments weathering out on the surface. Once fragments are found, you brush and scrape away the loose dirt to see if more bone is buried below. If so, and the specimen is deemed important enough to collect, a second operation called "quarrying," is begun to excavate the bones. As soon as they finished pitching their camp, Brown and Lull set out to prospect, and success in their quest quickly followed:

We are now camped seventeen miles from the Missouri River on the head of Hell Creek. Pulled in here to the old Sieber ranch Wed night. Dr. Lull and I started out Thursday morning and I located two good prospects one in clay which we have not worked. The other is a Triceratops or Torosaurus. . . . there may be plenty of fossils here for we have prospected only the one day. . . .

The country greatly resembles Lance Creek in Wyoming along the head of the breaks but the main canyons are certainly *bad* lands, almost impossible lands I might say . . . The great drawback to this region is the distance to freight fossils. It is over a hundred and thirty miles to Miles, the only available point.[20]

As the crew prospected for fossils throughout the rugged ridges and ravines, their first priority was to find the skeleton that had triggered their trip. Hornaday's *Triceratops* was quickly located and, although it was impressive, it simply served as a prelude to the unprecedented discovery that soon was to follow. On August 12 Brown wrote his boss in New York:

"Quarry No. 1 contains (several bones) of a large Carnivorous Dinosaur not described by Marsh . . . I have never seen anything like it from the Cretaceous. These bones are imbedded in flint-like blue sandstone concretions and require a great deal of labor to extricate."[21]

It was with this simple, matter-of-fact, brief mention that Brown announced the discovery that would catapult both him and the animal he found into the cult of celebrity. The bones he had written Osborn about belonged to the most intimidating carnivore yet found to have walked the earth, *Tyrannosaurus rex*. Brown immediately recognized the significance of the find. But he also realized that such a gargantuan skeleton would be difficult to excavate and transport 130 miles over the rut-riddled roads to the railhead at Miles City, and he was anxious to keep costs manageable in order to stay in good graces with Osborn:

> I purchased three good horses, a new wagon and camp equipage which will sell for nearly full value, about two hundred seventy five dollars.
>
> All provisions, lumber and plaster are very expensive so that I go as economical as possible. Plaster costs five dollars per barrel in Miles City so I use flour paste wherever practicable.[22]

Jordan's small frontier store was apparently ill-equipped to handle all the demands of its new paleontological customers. To quarry and collect a large fossil bone you must first dig around it, usually with picks, rock hammers, chisels, and awls, leaving a pedestal of matrix underneath the bone. Then you cover the fossil with a layer of protective material, such as newspaper or tissue, before applying more layers of plaster-soaked burlap to form a hard cast around it. Stout lumber and

Vertebrae of *Tyrannosaurus rex* as they were found in the quarry by Barnum Brown and his field crew in the badlands of the Hell Creek Formation along Big Dry Creek in 1908.
(AMERICAN MUSEUM OF NATURAL HISTORY LIBRARY, IMAGE #18337)

plaster were staples of fossil collecting, essential for constructing the sturdy cast around the bone to protect it during transport. It's a time-honored technique that even I have used in extracting more modest fossils from Hell Creek. But in Brown's day there weren't jackhammers and backhoes to help excavate large specimens. The bulk of the overburden entombing the skeleton had to be blasted away with dynamite, as Brown recounted on September 3:

We are still at work on quarry no. 1 [on the] Carnivorous
Dinosaur . . .

. . . the bones are separated by two or three feet of
soft sand usually and each bone is surrounded by the
hardest blue sandstone I ever tried to work in the form
of concretions.

There is no question but what this is the find of the
season so far for scientific importance . . . It is necessary
to shoot the bank and as it is not accessible to horses
work goes slow.[23]

With the snows of fall and winter approaching, Brown began
to have other members of the crew transport some of the fossils
back to Miles City, while he prospected for new fossils around
Hell Creek. Lull returned to New York, but Brown's continuing
efforts led to the discovery of four "crocodile" skeletons in the
beds above the dinosaur-bearing rocks. By the first of October
the weather was changing, but Brown found the change to be
"cool and exhilarating." Nonetheless, by October 13 Brown was
on his way home with a hefty treasure of priceless fossils:

Just arrived in Miles with rest of fossils. There are about
nineteen boxes in all, haven't finished packing, aggregat-
ing between ten and fifteen thousands of pounds weight.

Rate from Miles to New York is $3.13 per hundred
which at lowest tonnage means over three hundred dollars freight.[24]

Not even the cost-conscious Osborn quibbled. As the head of the museum's recently established Department of Vertebrate Paleontology, he desperately sought magnificent specimens for both exhibition and scientific study, and as it turned out, Brown's expenses were a piddling price to pay for what would become the most famous dinosaur of all time. Excluding shipping costs, total expenditures for the crew's travel, equipment, supplies, horses, and labor came to $1,345.

In the intervening years, as the tyrannosaur specimen was prepared and scientifically described, Osborn and Brown longed to see more. Doubts arose as to whether all the bones preserved in the quarry had actually been exposed and collected. So, in 1905 Brown once again set out for Hell Creek to expand the excavations at the tyrannosaur quarry. He almost didn't survive the trip out. On June 5 a relieved but exasperated Brown confided to Osborn that he had come face to face with an unexpected adversary on the sparsely populated plains:

I . . . am now on the road to Hell Creek.

The roads have been in an almost impossible condition for it rains two or three days at a time with one or two good days between.

We were out only three days when the [wagon] team [was] frightened by a motor cycle, pulled the picket pins and ran into a barbed wire fence. One was badly cut on the shoulder and cannot be used as a wagon horse for two or three weeks. Have been forced to put in one of the saddle horses to pull the load.[25]

Nonetheless, Brown finally made it to the same ranch that had served as his base of operations three years earlier. He learned that a competing crew of fossil collectors had visited the region earlier that year, but he was relieved to be able to report to Osborn that they had left before he arrived after not finding any fossils to their liking.

Complications were also arising with some of the local ranchers who, having witnessed Brown's determined search for skeletons in 1902, kept an eye out for potential prospects as they roamed their lands in Brown's absence. One family wrote the museum to offer it a specimen they had found for the "wild" sum of $8,000, but Brown recommended that contact with them should be limited to keep the museum from appearing too "anxious" to obtain it at such an exorbitant cost. Finally, by June 24 all those trials were put on the back burner, as work in the quarry was producing unexpectedly rich results:

> Yesterday we struck a concretion containing a large skull bone and today found another. . . . The large concretion weighs six or eight hundred pounds. . . . These concretions are at least six feet back in the bank from the ones taken out in 1902 and I had very little hopes of finding any more but now I will make a fifteen foot cut in the bank which will take at least three weeks with powder and horse power for it is a solid sand bank. . . .
>
> Please tell Mr. Hermann to send me ½ doz. short heavy chisels of best steel tempered for these hardest concretions also doz. crooked awls without handles. . . . [26]

Even without all the desired implements, which proved impossible to buy outside of Miles City, Brown's crew borrowed sufficient tools from the locals to uncover more bones. Much of the bulk excavation was again accomplished with the aid of dynamite and a horse-drawn scraper. But Brown also found time to prospect on a trip to Jordan, during which he found a "shattered" skull of *Triceratops*, which, despite the missing horns, could still be prepared at the museum to produce a "fine specimen." But the crew's main focus continued to be the tyrannosaur quarry, and by July 15 the crew had expanded the pit to enormous proportions, despite having to toil through heavily cemented sandstone in the torrid summer heat:

> Am now at work on the second large cut in the (tyrannosaur) quarry which is 100 ft long, 20 ft deep and 15 ft wide; hard sandstone which has to be blasted before it can be plowed. This is a heavy piece of work but (tyrannosaur) bones are so rare that it is worth the work.[27]

A full three weeks later Brown and his crew were still struggling in the quarry, and Brown was becoming concerned that the summer was rapidly coming to a close. The operation had been quite successful, but he was laboring under Osborn's quite reasonable direction to glean every bone possible from the concretionary outcrop. Thinking ahead, Brown began to plan for contingencies and seek Osborn's advice:

The work on Tyrannosaurus is being pushed rapidly. I have a force of three men besides myself and it is imperative that I supervise this work till completed, that is till this 20 ft cut is taken out and then if the bones continue into the bank as I think they will I shall let a contract to have the upper 20 ft of sand removed this winter for that will be cheaper than keeping an expensive outfit on the ground. There are about 600 cubic yards of sand stone above this cut which will cost 80 (cents) per cu. yd. and this figures out cheaper than removing it myself for labor is high here. I trust this meets your approval and I should like your opinion immediately. It means a great deal of expense but this is the rarest animal in the (Hell Creek) and I have never seen another fragment of it anywhere.[28]

Although the bones were occasionally difficult to identify in the field because of the cemented sandstone encasing them, Brown felt confident that he had found the other femur (or thigh bone), the humerus (or upper arm bone), an ilium (or hip bone), along with several ribs and fragmentary skull bones.

Finally, two weeks later, on August 22 Brown's excavation was nearing completion, after no more bones had been found leading into the cut bank. The crew had toiled under the scorching sun for two months, but these trials were not limited to just macho males. At that time Osborn was contemplating a trip out to Hell Creek in order to celebrate Brown's success and see the quarry for himself. This apparently raised an issue that

Brown had not anticipated, and he felt obliged to come clean about the identity of one of his crew members he had not told Osborn he was bringing. So he forthrightly fired off a preemptive defense:

> Mrs. Brown did accompany me from my home to camp and has done the cooking for the outfit this summer reducing our living expenses about half.
>
> I did not discuss the matter with you for it seemed a purely personal matter with me as long as I performed my duty without any added expense on her account and the Museum has certainly been the gainer.[29]

Although Brown was a pioneer in this regard, further social transformations followed that I heartily endorse. Coed field crews are now the order of the day at Hell Creek, but women's roles are far from limited to housekeeping. My female colleagues include both tenured professors and energetic students who specialize in disciplines ranging from paleobotany to theropod anatomy and microscopic eggshell structure. Many are expert collectors whose contributions to our projects have greatly expanded our knowledge of Hell Creek's fauna and flora—a result that Brown would have relished. Besides, most can drink with the best of us boys.

Toward the end of the season Brown refocused on the specimen for sale by the local ranchers. His strategy of feigning disinterest had already started to work. Brown had intentionally not taken time to personally inspect the specimen because of his "pressing" responsibilities at the tyrannosaur quarry. As a result, Brown reported to Osborn on August 22 that the asking price had precipitously dropped from $8,000 to the "low" price of $2,000, but Brown was determined to drive an even harder bargain for his boss.

In the succeeding month he inspected the specimen first-hand, as it lay in the ground, and confirmed that it was a duck-bill. The complete vertebral column was clearly visible, along with one femur and one side of the lower jaw. Brown was confident that the rest of the skull and limbs were also present below the surface, but he "thought it wise not to show up any more bones until I made the purchase." His final round of negotiations reduced the price down to $200, with an additional $50 due upon the start of excavation, which he planned for his next trip. In the meantime the specimen would be buried under sediment to protect it from discovery and weather-related damage. So, what turned out to be a nearly complete duckbill skeleton was secured for about 3 percent of the original asking price. Other trip expenses totaled $480. Brown was not only Osborn's best dinosaur collector but also his best purchasing agent.

By 1908 the initial scientific publications describing *Tyrannosaurus* had been published, but many parts of the animal remained unknown, and no one had an accurate picture of what the skull looked like. Both Osborn and Brown were determined to fill in the gaps, so another expedition was planned in which Brown would move his Hell Creek operation about thirty miles to the east along the Big Dry, down which Miles had chased Sitting Bull thirty years earlier. Brown had found a couple of prospects in that area in 1906 and was returning to evaluate those, as well as search for new skeletons.

Although his first few reports back to Osborn were not terribly encouraging, Brown and his colleague Peter Kaisen, an expert collector and veteran of the museum's early expeditions to Wyoming, pursued their quarry with dogged determination, interspersed with several lively social events at the local ranch. In addition to working hard, Brown seemed intent on enjoying himself and immediately set about sizing up his prospects. On July 8 Brown wrote Osborn that his paleontological luck had also taken a dramatic turn for the better:

> At last I have some good news to report. I made a ten strike last week finding fifteen [tail vertebrae] connected; running into soft sand. We have made a six foot cut with pick and shovel and found bones continued in sandstone concretions so have moved camp to the specimen and borrowed scraper and plow for a big cutting.[30]

Fortunately this specimen was not located in as precarious a setting as the original tyrannosaur quarry. Work progressed quickly under Brown's direction. Aided by the expert assistance of his experienced colleague Kaisen, Brown had more time to prospect for other specimens. Despite all this good news, however, all was not going smoothly in camp. Since his wife had not accompanied him on this trip, finding a trustworthy cook was proving to be quite a difficult challenge, and would continue to complicate the crew's activities throughout the season, regardless of Brown's expressed optimism in his letter of July 8:

> Have a very good man as a cook and teamster. Johnson by name. Our first cook from Miles found the work too heavy when it came to the pick and shovel so he left us last week.[31]

Nonetheless, the significance of the new skeleton had become abundantly clear by July 15. Although the bones in the new quarry were partially imbedded in sandstone, the rock was easily chipped away. Within a week enough diagnostic parts of the bones were exposed for Brown to make a positive identification, and he gleefully exclaimed to Osborn:

> Our new animal turns out to be a Tyrannosaurus . . . We have the complete vertebral column except the tip of tail, skull and lower jaws, complete pelvic girdle in position and the (neck) ribs . . . so far no limbs have appeared . . . The bones are in a good state of preservation . . . [32]

At long last Osborn and Brown could revel in viewing almost the whole animal; the specimen included the animal's four-foot-long skull, studded with six-inch-long, serrated teeth.

Most of the skull for the skeleton found in 1902 had not been found. So, in his July 30 response, Osborn was ecstatic.

> Your letter of July 15 makes me feel like a prophet and the son of a prophet, as I felt that you would surely find a *Tyrannosaurus* this season . . . I congratulate you with all my heart on this splendid discovery . . . I am keeping very quiet about this discovery because I do not want to see a rush into the country where you are working.[33]

The scientific agenda for the expedition was rapidly moving toward a historically successful conclusion, but the domestic arrangements within the camp were far from ideal. Retaining a reliable cook continued to bedevil Brown, as he lamented on August 1:

> I had everything running well in camp but our cook drove the team in yesterday and today he was so drunk that I let him go. Was sorry to lose him for he was a good cook and a good worker but I won't have a man in camp that I cannot trust to town with the bones. Hope we may have someone who can cook without burning the water when you come out.[34]

Osborn, who was again contemplating a trip to the site to see the marvelous specimen for himself, playfully responded:

I hope . . . you will be able to discover a fairly good cook as I know I shall enjoy the badlands much more if the water is not burnt.[35]

On August 10, in anticipation of Osborn's arrival and first-hand inspections of his new discovery, Brown confidently stated:

I am sure you will be more pleased with our new Tyrannosaurus when you see what a magnificent specimen it is. . . . This skull alone is worth the summer's work for it is perfect. We finished boxing it and moving it out of the quarry Saturday. This block weighed about [3000 pounds] while the lower jaws weigh about [1000] lbs.[36]

Osborn arrived on August 26 and apparently spent about a week examining the quarry and prospecting with Brown. Their biggest problem involved getting the heavy casts containing the tyrannosaur bones out of the quarry and across to the north bank of the Missouri. There the nearest rail link was located in Glasgow, northwest of Fort Peck—where Miles and Sitting Bull had obtained their supplies—and forty-five miles from Brown's camp. The transport would require the help of several extra men and horse-drawn wagons. Brown wanted to ship the whole collection at one time and load it on a train going straight through to New York, so the boxes and bones would not be damaged while being transferred between different trains. Most of September and the first week of October involved building boxes for the bone-filled casts and packing the specimens in the crates. The weather was quickly deteriorating, and by the last week of September and the first week of October storms stocked with a mixture of rain and snow arrived to complicate Brown's efforts. Getting the boxes from the quarry to the Willis ranch, which Brown used as his base of operations, proved to be the weak link, because there were no established trails through that section of badlands. Twice the wagon got stuck, to the point where Brown had to abandon it and round up extra horses to pull it out. But, on October 8 Brown departed for Glasgow with a convoy carrying five loads of fossils pulled by sixteen horses, which covered the forty-five miles without incident by the end of the next day. On October 10 the boxes were loaded on the train, which began the journey to New York on the twelfth, where they safely arrived. The costs of this expedition to secure the first skull of *Tyrannosaurus* totaled $1,171.

But even then the perils surrounding these first *Tyrannosaurus* specimens were not over. During the outbreak of World War II the American Museum of Natural History sold one of the skeletons to the Carnegie Museum in Pittsburgh. As Brown wrote,

> The [first] specimen which had most of the limb bones preserved was sold . . . after we had made casts of the limbs in 1941; as we were afraid the Germans might bomb the American Museum in New York as a war measure, and we had hoped that at least one specimen would be preserved.[37]

Skeleton of *Tyrannosaurus rex* as it was originally mounted in an upright posture for exhibit at the American Museum of Natural History. (AMERICAN MUSEUM OF NATURAL HISTORY LIBRARY, IMAGE #315110)

Fortunately both skeletons survived, but I have a closer affinity to the second. For that was the specimen that my staff remounted during the renovation of the fossil halls in the 1990s. It was a daunting assignment, since each bone had to be removed from the old upright mount, reconserved, and remounted in the new, more animated posture prescribed by recent research. It took two years to accomplish, a period replete with unremitting worries over the welfare of this priceless specimen. But our crew did a spectacular job, and Brown's skeleton stands ready to pounce on its prey in the renovated hall.

Brown went on to mount other spectacular expeditions north of the Canadian border along the Red Deer River. This region had also been of great interest to Lewis and Clark, because Thomas Jefferson had hoped they would discover tributaries flowing south into the Missouri that would extend the boundaries of the Louisiana Purchase and provide access to the fertile prairies and rich, fur-trapping landscapes of the northern Rockies. As it turned out, such rivers did not exist, and the area remained isolated and remote for more than a century after Lewis and Clark had passed.

Between 1910 and 1915 roads in that region were either nonexistent or impassable. So Brown devised a completely new way to prospect and collect the dinosaur skeletons exposed by erosion on the steep bluffs along the river course. Seeming to take inspiration from Lewis and Clark, the crew constructed a flat-bottomed boat measuring four yards wide and ten yards long. A tent pitched on deck served as the kitchen as the crew floated down the river, stopping to prospect whenever promis-

ing outcrops came into view. The fossils were stored on the flat-
boat until the end of the season's trip, when they were unloaded
and shipped back to New York.

But despite the romantic image that these descriptions
conjure up—lazily floating downriver, plucking magnificent
dinosaur skeletons from the banks—the reality was often quite
different. Photos show the crew with their heads covered with
masks of cheesecloth. Brown revealed the itchy derivation of
this field innovation as follows,

Barnum Brown's field crew from the American Museum of Natural History on their flat-
boat with their mosquito-repellent headgear as they collected along the Red Deer River in
Alberta, Canada. (AMERICAN MUSEUM OF NATURAL HISTORY LIBRARY, IMAGE #18547)

It has rained all over this part of Canada as never before—the Red Deer River is out of banks most of the time; ten feet of water was running over our last year's camp site and came to where our fossils were parked last year. Mosquitoes are fearless of smoke, ferocious and in numbers equal to the Kaiser's army.[38]

These early paleontological explorers were also devotees of Dionysus. They intertwined the tiring travails on their collecting expeditions with lively parties and drinking sessions with the locals. Despite all the hardships, Brown filled his days and nights with more than hard labor and battles with mosquitoes, as he roamed the western badlands in search of fossils. His field wardrobe contained a splendid beaver skin coat, which archival photos show him sporting even on the outcrops. Such duds probably came in handy on special evenings when the field crew would gather with the locals for celebrations like the following one at a ranch near Hell Creek that Brown recounts in his field notes.

In those days it was not all work on the ranches, or in fossil camps. We had a dance once at the Twitchell ranch— a 3 roomed log house. The music was a phonograph, and Bess Willis rode all over the country collecting records. She was the only unmarried female in the region. Babies were ricked along the floor on bed rolls, and as the women played out they took to the beds, and the men retired to the haymow of the barn. This dance was a gala 4th of July affair, starting in the afternoon of Friday, continuing all that night; and a few women lasted till Saturday morning, or recovered after sleep, to keep it going. Bess Willis went all the way through, for it was her party.[39]

I can assure you that parties are still staples of fieldwork in modern Montana. I've tried my hand at two-stepping to local country and western bands at Jordan's street dance on the

Barnum Brown in his legendary beaver-skin coat. (AMERICAN MUSEUM OF NATURAL HISTORY LIBRARY, IMAGE #19508)

fourth. The revelry reigns until dawn, and the fifth involves licking your wounds. I've also celebrated the fourth with members of the same Twitchell family on the nearby ranch where the family moved when the dam at Fort Peck was constructed, flooding the old family spread. Our fireworks may not have been as thunderous as Brown's blasts in the quarry, but I know the results were as satisfying.

What might not be so clear from Brown's description is that such parties, along with other socializing in the bars with the locals, often serves an important paleontological purpose, as well as providing some much needed R and R. If we fast-forward several decades into the 1960s I can illustrate the point by introducing you to an expert dinosaur collector whom you may well have never heard of.

7

FOLLOWING IN THE
FOOTSTEPS

Flat out, the best fossil collector I have ever worked with, and the heir to Barnum Brown's legacy, is Harley Garbani. Harley is not nearly as well known to the public as several other contemporary paleontologists and collectors; however, his prowess as a collector has made him a living legend within the paleontological community.

In this age of rampant specialization, it might surprise you to learn that Harley was not formally trained as a paleontologist. Born in Los Angeles, Harley spent most of his youth on a "dry land" ranch in Southern California's San Jacinto Valley, where he actually trained and worked as a plumber. But early on he had already caught the bug, for he had found his first fossil when he was eight in the rocks near his home. In fact, it was part of a camel leg, which is not surprising, given the fact that the deserts of Southern California are laced with faulted and fractured rock units that formed during the Age of Mammals that began 65 million years ago. Nonetheless, the die had been cast for Harley's fateful date with dinosaurs.

Like Brown, Harley's paleontological fame derives from discoveries at Hell Creek, where he found the best specimen of *Tyrannosaurus rex* ever recovered up through the 1960s. It now forms a cornerstone of the collection at the Natural History Museum of Los Angeles County.

In 1965 that museum enlisted Harley to search for another specimen of *Tyrannosaurus*, although the museum's charge included the collection of other dinosaurs, such as *Triceratops* and duckbills. At that time the only two known specimens that were anywhere near complete had been found by Barnum Brown in the first decade of the 1900s, so it had been more than fifty years since a meaningful bundle of bones from the beast had been uncovered.

Harley approached this quest by quietly drawing on a set of personal traits, tempered by years of fieldwork in the desolate deserts of southern California and northern Mexico. On the hunt his quiver contained an arsenal of assets, including acute observation, enormous experience, enduring patience, sustaining passion, and disarming charm. That last characteristic might seem somewhat superfluous to his objective, but it proved essential to Harley's success, as the strategy that led to his most famous discovery will illustrate.

Harley's attitude in accepting this challenge reflected his uncommon blend of realism and common sense. As he intimated to me later, "I thought it was a long shot, but I had everything to gain and nothing to lose by trying." Undaunted, Harley set off with his wife and daughter for the same rugged badlands that Brown had made famous.

The now-familiar exposures of 65-million-year-old sediments at Hell Creek form a vast expanse of breaks along the southern flank of the Missouri River. For Harley that represented both good news and bad. On the one hand there were plenty of places to quest for his quarry. But despite the fact that *Tyrannosaurus* represented the largest carnivorous dinosaur yet found, stretching to a length of more than thirty feet, finding fragments of its skeleton weathering out of the labyrinth of badlands was tantamount to searching for the Minotaur in the maze.

Rather than wandering off blindly through the breaks, Harley decided to hedge his bet by unleashing his formidable talents for both drinking and socializing on the unsuspecting inhabitants of Jordan. As you might suspect by now, these roughly hewn ranchers of cattle and sheep are laden with streaks of defiance and independence as wide as the "Big Sky" country itself. Although friendly and generous once mutual trust and respect are established, they often meet outsiders with a healthy dose of skepticism.

To overcome this innate reticence, Harley made the rounds about town, buying supplies and setting up a checking account at the local bank to fund his efforts. He freely admits that he was "a babe in the woods" in terms of his knowledge of the area and its inhabitants, but while making his rounds with the merchants he openly discussed the reason for his presence and quest. That first year Harley came across only three people in town who were interested in fossils: a teller at the bank, the owner of the dry goods store, and an innately intelligent but soft-spoken rancher named Frank McKeever. Frank owned a spread about twenty-five miles northwest of town that included the rugged exposures along the gorge of Billy Creek, a tributary of Snow Creek.

Through the good offices of the bank teller, a time was arranged for Harley to meet Frank at a cabin in town. As Harley entered, Frank sat in a large easy chair and peered at Harley with an intimidating and discerning eye. Somewhat sheepishly, Harley explained his quest, and as if to short-circuit any further small talk, Frank immediately invited Harley to "follow me out to the ranch."

Frank's ranch near Snow Creek encompassed an area where his family had already lived for more than a generation. Frank's father had migrated into the Hell Creek area from South Dakota in 1914, at the vanguard of the wave of homesteaders flooding in from the East. He had just finished a short course in agriculture at South Dakota College, and he was anxious to try and put his new knowledge to use. In 1917, after marrying, he moved out to Snow Creek to establish his spread, and soon found himself supporting a small family that included Frank and his sister. Following in his family's footsteps, Frank was well educated in his own right, especially in his knowledge of the land. In addition, he was a native Montanan who had lived his whole life in the region. He was sure that he could help

Harley find the kind of quarry that he sought in the wilderness of breaks that riddled his ranch.

With Frank as his newly found patron, Harley headed out to the ranch after the July Fourth festivities in town, but before he got there Hell Creek taught him a valuable logistical lesson. A typical summer thunderstorm enveloped the landscape, turning the normally dusty road into a quagmire. Slipping slowly along through the slime, it seemed like the mud rather than Harley was driving the vehicle. Fortunately he ran into some fishermen returning to town. Stopping and sensing Harley's dilemma, they gave him advice that would serve him well in his quest, "You've just got to keep movin', or the mud will ball up on your tires and you'll be stuck here 'til it dries out." Accordingly, Harley and his intimidated family forged on through the gumbo to McKeever's.

Indeed, Harley's family did experience success that very first summer. They found a nice specimen of *Triceratops* weathering out of the blistering rocks. Harley had established a foothold in the unforgiving landscape and tasted a morsel of success, but he had bigger fish to fry.

Harley set up shop at the local watering hole, appropriately named the Hell Creek Bar, and began the process of introducing himself to other locals by buying drinks for the thirsty denizens. The protocol at the Hell Creek Bar is disarmingly if dangerously simple. When you walk in you buy a round for your friends, and in return they all buy one for you. Before long everyone assumes the role of long lost friends and trusted confidants, which was exactly what Harley was banking on.

Although one of the most gentle and peace-loving men I

have ever met, Harley walked in armed to the teeth—or rather with teeth. In his pocket he always carried some isolated claws and tooth fragments of *Tyrannosaurus*. Invariably, after a few drinks the brethren would inquire as to what Harley was doing in a parched place like Jordan. So Harley would pull the fossils out of his pocket and ask if they'd seen strange stones like those on their ranches. One night the strategy paid off when a crusty yet intelligent rancher named Lester Engdahl, whom you've already met, said he thought he had. He invited Harley out to his spread to take a look.

Yet, despite Lester's help two summers passed without cornering a tyrannosaur; nonetheless, Harley's efforts were far from fruitless. As he recounts:

> I'd collect every fossil from mice to *T. rex*. I just loved beatin' the bush.[1]

"Mice" is his nickname for the small ancient mammals that were fossilized along with the dinosaurs.

But Harley's bargain in the bar with Lester forever altered his paleontological fate. On July 27, 1966, he spied an enormous toe bone protruding out of a dark gray mudstone as he swung down a ravine below a small livestock pond, prophetically named Pearl, on the Engdahl ranch. It was broken in half, and he could immediately see that the bone was essentially hollow inside. His heart raced as adrenaline surged through his veins; for, hollow bones are a diagnostic evolutionary feature of meat-eating dinosaurs called theropods. A little farther along

Harley Garbani holding one of the lower jaws of the *Tyrannosaurus rex* skeleton he discovered and collected on the Engdahl Ranch north of Jordan, Montana, in the late 1960s.
(COURTESY OF LUIS CHIAPPE AND THE NATURAL HISTORY MUSEUM OF LOS ANGELES COUNTY)

the bank he saw part of a large hind limb bone beginning to weather out of the mudstone. One of the challenges of collecting in the field is to know enough about the anatomy of different animals to make at least a preliminary identification of what you have found. Size is certainly one clue, but their detailed shapes, especially of the ends of the bones where they fit up against adjacent bones, are another key. As Harley wrote in a letter to me on May 19, 2003:

By that time I was well aware of the difference between the foot and limb bones of meat and vegetable eating dinosaurs. The pieces were all eroding out of the same matrix. My notes don't give up the thrill that I felt when thinking at the time, "This could well be my *rex*."[2]

Although Harley perpetually refers to this priceless skeleton as "the damn thing," he quickly realized that, because of the hollow bones and their large size, he had come face to face with his long-buried treasure:

I was pretty excited. I didn't figure another of those suckers would be found. I'd seen enough of *Tracodon* (an old discarded name for a genus of duckbilled dinosaur) to know this was something bigger.[3]

Understated as usual, Harley simply inscribed the events of the day in his field notebook, much as Barnum Brown had done sixty years earlier:

Tried to shower several times during the night, but not enough to settle the dust. I spent all day on the Engdahl ranch, just west of the Trumbo Ranch fence. . . . In late P. M. went to new dam. . . . A short distance NE of dam in some small roughs, I found what I believe to be the large limb element, plus tarsal & 2 toe elements, of a Rex. Very large. . . .
  HJG 339
  HJGV666

Probable limb & foot elements of
Tyrannosaurus Rex
Hell Cr. Formation[4]   HJG

It's amusing that Harley's field number for the tyrannosaur locality turned out to be 666, "the sign of the beast" in the Book of Revelations. For this would prove to be a most impressive beast indeed.

Over the next three years Harley enlisted the aid of local volunteers to excavate the other bones of the body buried in the badlands. In one critical way, however, the operation remained a family affair between Harley and the Engdahls. Tons of gray silt and clay entombed the specimen in several feet of overburden, and all of it had to be removed delicately in order to get at the priceless skeleton. Unlike Brown, who used dynamite to blast away the overburden, Harley relied on both natural weathering processes and the heavy equipment skills of another of Lester and Cora's sons, Larry.

Yet, even with Larry Engdahl's bulldozer, the job was anything but easy. The unweathered mudstone that entombed the skeleton was too hard for even the bulldozer's blade to cut through, and in the process of his trying, Hell Creek unveiled another surprise for Harley. As Larry made a pass across the more weathered mudstone on the top of the bank that held the tyrannosaur, Harley noticed other bones and teeth strewn across the surface of the cut, as he confided to me in a conversation at the Hell Creek Bar in 2003:

These teeth were shorter and more slender than *rex* teeth. That moment was both heartbreaking and exhilarating. We had inadvertently cut through the top of an albertosaur skeleton, lying about two feet above the tyrannosaur skeleton. All I could do is collect all the pieces and carefully excavate the complete hind limb.[5]

In evolutionary terms, *Albertosaurus* is essentially the little brother of *Tyrannosaurus*. In terms of modern carnivores, albertosaurs are to tyrannosaurs as American mountain lions are to African lions; all are intimidating predators. Harley's "mother lode" of theropods on the Engdahl ranch remains, to this day, the most remarkable single discovery in the history of dinosaur paleontology, as far as I'm concerned. But, chastened by the inadvertent damage done to the albertosaur skeleton, Harley realized that patience was a key to collecting the underlying tyrannosaur specimen, for, as Harley states in his letter of May 19:

It took many seasons of dozer work to get down to the level [of the tyrannosaur skeleton]; had to let "Mother Nature's" winter weather break down the hard clay. Not till August 27, '68 after the fine dozer work of Larry Engdahl, were we able to expose the beautiful skull and jaw elements. What a sight to view those beautiful teeth in complete jaws.

It took a few more seasons before we finally cleaned up the area.[6]

In all, about 30 percent of the skeleton was preserved. That included almost 75 percent of the spectacular skull, studded with a full arsenal of sinister, serrated teeth for slashing the flesh of its unfortunate prey. The skull alone was nearly five feet long and the bones were not distorted in any way, unlike those in the skull that Brown found.

In short, that's the story behind how the third and most complete skeleton of *Tyrannosaurus* known at the time was discovered. We professional paleontologists really can't take any of the credit.

Harley's success ignited renewed paleontological interest in the region surrounding Hell Creek. Bill Clemens had worked in Wyoming while conducting research on the minuscule mammals that lived alongside the tyrannosaurs, and he was anxious to see if the fossils he found in Wyoming were similar to those around Jordan. He got his opportunity when Dave Whistler, then Chairman of Los Angeles County's Department of Vertebrate Paleontology, called him. As Bill related to me through an email in April of 2003:

Dave Whistler had been interested in the Cretaceous mammals that Harley and his crews had discovered, but had plenty of other things to do so never really followed up on their discoveries. When Harley brought in Paleocene mammal material Dave recognized the research potential. Dave called me and asked me if I would be interested in following up on this and Harley's earlier discoveries of Hell Creek mammals. Dave's invitation resulted in my joining Harley in Garfield County. That summer and for a couple of summers thereafter we ran joint expeditions. LACM funded Harley and got the big dinosaurian material we collected and UCMP [the paleontology museum at Berkeley] supported our work on the mammals. Needless to say I and all the UCMP crews owe a significant debt of gratitude to Dave for his invitation.[7]

After the museum in Los Angeles completed its project, Harley signed on with Bill Clemens of Berkeley to continue his search. Throughout the late 1970s and early 1980s Harley and Bill teamed up to "beat the bush," with crews of Berkeley students wandering in their wake. They roamed throughout the Engdahl ranch, as well as several others, including the Twitchell ranch near where Brown had found his skull. Bill and Harley uncovered more fragmentary remains of *Tyrannosaurus*, including part of a massive upper jaw in 1977 and additional upper and lower jaws in 1982.

Their efforts also greatly expanded our knowledge of the smaller animals that filled out the ancient environment, and Harley played a pivotal role in that quest. Then nearing seventy, Harley still possessed the superhuman eyes and sagacious instincts of an owl when it came to finding fossils, and not just large dinosaur skeletons. Harley also excelled at finding the almost microscopic teeth of the ancient mammals that lived in the dinosaurs' shadows. He accomplishes this feat with the aid of his beloved "cheaters," a pair of bulky gray magnifying goggles, used by jewelers and gemologists, which he dons, along with knee pads, to crawl along the ground and examine small areas where bones accumulated in the ancient stream beds.

One day, when I was a student, I came upon Harley in the field. I thought he had lost his mind. He was kneeling next to an eighteen-inch-tall anthill, intensely staring through his "cheaters" at the pebbles on its side. He explained that ants sometimes use the small fossilized teeth along with pebbles in order to fortify the side of their nest against rain and erosion. As he held out his palm, I saw that he had harvested several small teeth from the anthill's fortifications. "You've got to be quick. They're not altogether pleased by your presence and can deliver a pretty good bite," he emphasized, as the ants scurried out in defense of their palace.

As Harley's nickname for them implied, most of these mammals were no larger than modern rodents. The dinosaurs that ruled their domain dwarfed them. Nonetheless, based on the cusp patterns found on their teeth, many have evolutionary links to modern lineages of mammals, including insectivores, many carnivores, and hoofed herbivores. Yet, none achieved large size until after large dinosaurs perished.

The tyrannosaur that Harley and the Engdahls unearthed helped catalyze my own career in 1969, when I was a volunteer at the Natural History Museum of Los Angeles County. Although eight or nine years older than I, Dave Whistler, then head of the Vertebrate Paleontology Department, shared a connection with me. My father was Dave's high school principal in our small hometown outside Los Angeles. The summer after I graduated Dave graciously took me under his wing as a summer volunteer, after my father explained my interest in paleontology. One of the tasks I undertook was to help get Harley's tyrannosaur jaw x-rayed. Tyrannosaurs, after all, were reptiles, and other reptiles continually replace their teeth throughout their life. So, Dave and the other curators wanted to see if unerupted teeth were hidden inside the jaw. The scene was a bit surreal, as we wheeled the four-foot-long jaw on a gurney through the UCLA Medical Center to the Radiation Lab. Although we made sure that the tyrannosaur's jaw did not drop to the floor, the jaws of many bystanders certainly did. But, to my amazement, after an hour of photographing and developing, the X rays revealed numerous, scythelike teeth still waiting to erupt and terrorize the tyrannosaur's prey. This brush with cutting-edge science served to cement my commitment to a career in paleontology and with Dave's assistance, I went on to study at UC Riverside and Berkeley.

It was as a student at Berkeley, during the 1979 field season, that I was initiated at Hell Creek under Bill and Harley's tutelage. When Harley was asked how he found three of the first five known tyrannosaurs, he humbly replied, "I'm just a pretty lucky fella. They're so rare, it's just a privilege to find them,"[8] but I didn't believe that for a second.

Serendipity does, almost always, play its role in great paleontological discoveries, but serendipity requires much more than good luck. Harley assiduously employed all his personal and professional talents in the pursuit of his tyrannosaur. He thoroughly earned his success with dogged determination and relentless passion throughout his pursuit. In fact, as he once told me, "It's the passion that's most important. Without passion to achieve your goals there is little reason for living." That passion for discovery sustained Harley in an unfamiliar land inhabited by intensely independent people whose help he desperately needed to succeed. As he tells all of us students who work with him, "It's very important to nurture strong and trusting relationships with the locals. You can't just take the fossils you find and leave town; you might well need their help some day." Both Harley and Bill continually emphasize that lesson, and it has served me well wherever I have traveled on expeditions. Harley's discoveries at Hell Creek were truly Herculean accomplishments, and the privilege is mine to have joined him and toasted his deeds in the draws of those daunting badlands.

8

FATAL IMPACT AND ENORMOUS ERUPTIONS

Throughout my first year at Berkeley things went according to plan, as I expanded the geologic mapping of the Hell Creek area in preparation for writing my dissertation under Bill Clemens's guidance. But those well-laid schemes were dashed during my second year when an earth-shattering hypothesis blew out of Berkeley's geology department, one floor above my office. Its reverberations are still roiling the scientific community.

The source of the detonation derived its power from the research of a geologist named Walter Alvarez. Along with his Nobel-laureate father, the physicist Luis, Walter proposed that the extinction of large dinosaurs resulted from the impact of a six-mile-wide asteroid with the Earth.[1] Their initial evidence for this debacle was contained in a one-inch-thick layer of clay separating the marine rock layers that contain Cretaceous and Tertiary fossils at Gubbio, Italy. After chemically analyzing the clay, the Alvarez team noted that it is highly enriched in an element called iridium, which is rare in the rocks of the Earth's crust but more abundant in meteorites.

The impact hypothesis quickly catalyzed a determined debate concerning the cause of dinosaur extinction at the end of the Cretaceous Period, 65 million years ago. In essence, that debate pits the Alvarez hypothesis against a more conventional, earthly scenario, which claims that the extinction resulted from a massive pulse of volcanic activity long-recognized to have occurred at about the same time. This volcanism formed the extensive geologic province in India now called the Deccan Traps.

In order to help envision these conflicting scenarios, let us descend once again to the depths of Hell Creek and imagine what those scenes might have looked like on that broad, forested floodplain. First let's peer back in time and examine how the more gradual extinction scenario—based on volcanism—might have played out.[2]

The last summer day of the Cretaceous dawned like many others over the previous half million years. The atmosphere was thick from ash and toxic gases spewed out of volcanoes in the rising ancestral Rocky Mountains to the west, as well as from widespread volcanic eruptions in India, the South Atlantic, and other areas around the globe. Hephaestus, the volcanic god of classical mythology, was putting in serious overtime. In India alone these massive eruptions produced a flood of about 480,000 cubic miles of black basaltic lava, making this volcanic event one of the largest in the 4.5-billion-year-history of the

Earth. These flows of lava, piled one on top of the other, formed a blanket over one and a half miles thick. In total they covered an area larger than present-day California. Trillions of tons of toxic pollutants, such as carbon dioxide, sulfur, and chlorine, were injected into the atmosphere.

Over time, as the result of deteriorating environmental conditions across the coastal plain and around the globe, the once-thriving populations of horned dinosaurs such as *Triceratops* dwindled. So did those of duckbills, such as *Anatotitan*, and their predatory nemesis, *Tyrannosaurus*.

The flora and fauna living along the lush coastal floodplain in today's Hell Creek region were subjected to endless episodes of corrosive acid rain, catalyzed by the air pollution from the eruptions. As if the ash and toxic gases weren't enough, temperatures changed too. For hundreds of generations the shallow seaway to the east of the floodplain acted like an environmental cushion, helping to keep the temperature and humidity in a tepid, subtropical range that provided the dinosaurs with abundant water and food. But a million years ago the shoreline began to recede back off the continent and return to positions along the major ocean basins to the north and south. Over the last 50,000 years that retreat had accelerated. Thus, days became hotter and nights became colder; summers grew warmer, and winters grew cooler. Rates of precipitation also changed. During the warmer times of the year the temperature rose to uncomfortable levels as the result of global warming generated by the volcanic particulates and aerosols.

These environmental changes annihilated the herds of duckbills, and in turn extinguished the tyrannosaurs that preyed on them. But on this day 65 million years ago, the last *Triceratops*, weakened by the extreme heat of the noonday sun and the choking dust from last night's eruption, slumped to the ground and expired. A lineage that had ruled the continents for over 150 million years had finally perished.

Now try to envision that very same coastal floodplain at Hell Creek, assuming the more catastrophic impact scenario proposed by the Alvarezes occurred.

The summer night was calm and pleasant, punctuated only by the occasional grunts and snorts from a slumbering herd of *Triceratops*. Through the stand of conifers adjacent to the herd a bull tyrannosaur slowly and intently stalked toward a plump and peacefully sleeping juvenile near the edge of the unsuspecting herd.

Just as the bull edged within striking distance, a meteor many times brighter than the sun ignited the southern sky. The meteor skimmed across the heavens as if hurled by Zeus himself, streaking at somewhere between 50,000 and 150,000 miles per hour and ripping through Earth's atmosphere in a matter of seconds. The whole herd awakened instantly and stampeded in startled panic toward the north, ignoring the tyrannosaur fleeing in their midst.

To the south the Earth was rocked by the shockwave from

Artist's depiction of the meteorite impact off Yucatan at the end of the Cretaceous Period about 65 million years ago, which many scientists believe played a major role in the extinction of large dinosaurs. (PRINTED WITH PERMISSION FROM THE ARTIST, WILLIAM K. HARTMANN)

the enormous impact near present-day Yucatan. The six-mile-wide meteorite slammed into the shoreline of the Gulf of Mexico, generating temperatures of several thousand degrees as the incandescent object excavated a crater between 50 and 60 miles across and between 13 and 25 miles deep. The impact generated an earthquake of magnitude 13—one million times greater than the strongest earthquake ever recorded in human history. The explosion was equivalent to simultaneously detonating 10,000 times the number of nuclear weapons contained in all the world's arsenals at the peak of the Cold War. Five thousand cubic miles of material were blasted out of the crater and launched into orbit at a velocity equal to 50 times the speed of sound. This was 1,200 times the amount of material that erupted out of Mount Saint Helens in 1980. Part of the Earth's protective atmosphere was literally blasted away by the ejected debris.

The force of the concussion generated tsunamis 300 feet high. These "tidal waves" instantly decimated organisms living nearby. Within hours, the tsunamis struck other shorelines around the Gulf of Mexico, eradicating near-shore communities living hundreds and even thousands of miles from the point of impact.

As orbiting particles reentered the remaining atmosphere, they again heated the air to Hadean levels comparable to "broiling" temperatures in your oven. Wildfires ignited across several areas of the globe. Even at a distance of several thousand miles, the environment at Hell Creek was severely stressed.

Over the next several months after this incineration the atmosphere became so choked with dust that no sunlight penetrated to the ground. Without the warmth from rays of the sun, temperatures plummeted below freezing for at least a month, and maybe as many as six. The earth's flora foundered. The surviving plants on which the herbivorous dinosaurs fed soon died, and with the death of their food source the *Triceratops* herds perished. Deprived of their prey, the stunned tryannosaurs followed suit as the dusty curtain fell on the Cretaceous.

. . .

Both casual hypotheses were compelling and riveting to a new-comer like me. But as a scientist and his student, Bill Clemens and I had to evaluate the evidence contained in the rocky chronicles of Hell Creek's sedimentary layers to see which scenario seemed most plausible.

Initially the impact hypothesis was met with considerable skepticism, because geological research has long been directed by a simple dictum: In interpreting geologic history, "the present is the key to the past." In other words, when one seeks to interpret

The participants in the dinosaur extinction seminar at UC Berkeley in 1980. Seated from left to right are Kevin Steward, Helen Michel, and Dale Russell. Standing from left to right are Lowell Dingus, Alessandro Montannari, Luis Alvarez, Frank Asaro, William Clemens, Walter Alvarez, and Mike Greenwald. (PHOTO BY SAXON DONNELLY COURTESY OF THE CAL MONTHLY AND THE UC BERKELEY ALUMNI ASSOCIATION)

how ancient layers of rock formed, one need only look at events that form rocks in our modern environments. For example, rivers deposit layers of sand and mud that eventually harden into sandstone and mudstone, like those that form the badlands throughout the Hell Creek region. Volcanic eruptions spew out lava flows that eventually harden into volcanic rocks, and so forth. Although the Earth is pocked by a few large craters resulting from impacts of space debris, few clear examples are apparent, and such impacts during recent geologic history are thought to be relatively rare. Thus, since we don't see large impacts occurring frequently in our world today, the idea that an immense impact caused the extinction of the dinosaurs seemed more like a scenario out of science fiction than scientific research. Conventional scientific explanations for the ex-tinction relate to more widely accepted geologic processes, such as extensive volcanic activity and the changing configurations of oceans and continents— both of which could cause severe environmental and climatic perturbations that might have triggered the extinction.

In a sense, the asteroid that the Alvarez team envisioned landed squarely in my lap. The field area where Bill and I worked contains one of the world's most complete records of the evolutionary changes that occurred on the continents at the end of the Age of Dinosaurs and the onset of the Age of Mammals. If an impact extinguished the dinosaurs, there might well be some evidence for it right there at Hell Creek.

Although Bill and I were decidedly skeptical of the impact scenario, a series of seminars was scheduled with the Alvarez team to explore the evidence and its implications. Initially these encounters were quite collegial, and both sides seemed to revel in the interdisciplinary exchange of information and ideas. Luis, never at a loss for ego as well as intense intellect, confidently stated that he'd make believers out of us within six weeks. It seemed to be his perspective that he and Walter were just helping us solve a thorny evolutionary puzzle that had bedeviled paleontologists for generations.

The main outgrowth of these early seminars was the establishment of a field program that I helped Bill conduct the following summer at Hell Creek. In essence, we looked for a clay layer enriched in iridium similar to the one that Walter found in Italy. Basically, this entailed collecting rock samples every three or four inches from our sequence of rock layers that spans the boundary between the underlying layers, which contain dinosaur fossils, and the overlying layers representing the beginning of the Age of Mammals, which do not. These samples were then turned over to the Alvarez team for chemical analysis. It was rather tedious work, carried out under the blazing Montana sun during the infernal months of July and August.

Despite our skepticism that any anomalously high concentrations of iridium would be found, we were dead wrong. One clay layer very close to where the Cretaceous–Tertiary boundary is marked by changes in the ancient flora contained a significantly higher concentration of iridium than the other layers in our sequence.

The Alvarez team immediately declared victory in both

Outcrop along the road to Hell Creek State Park showing the iridium-rich clay layer (light gray stripe near the middle of the black coal bed) that contains the fallout from the impact that occurred at the end of the Cretaceous Period about 65 million years ago. (LOWELL DINGUS)

our seminars and in a series of debates that Bill, Walter, and Luis held at Berkeley and other scientific conferences. But Bill was not interested in conceding. Especially troubling to him was the fact that, across our field area there was a gap of about ten feet between the highest dinosaur skeleton we could find and the clay layer that contained the iridium. Since lower layers are distinctly older than higher layers, this represented powerful evidence to Bill that the last dinosaur had expired well before the impact, or whatever other event caused the iridium enrichment. Walter and Luis, on the other hand, felt that this gap was simply a statistical artifact. In other words, given the sparse distribution of skeletons found in the Hell Creek Formation, one could not necessarily expect to find a skeleton right at the boundary.

These and numerous other contentious issues quickly began to inject our seminars and debates with more personal venom. Numerous other scientists jumped in to add their two cents on subjects ranging from astrophysics to geochemistry to physiology to volcanism. Eventually, the debate seemed to center on whether the iridium in the clay resulted from the impact of an extraterrestrial object or from volcanic activity. In my own mind, I came to see the debate as the "Iridium War," although both intellectual and political battles are being fought on numerous other fronts.

Whereas Walter seemed to take the debate less personally and accept the whole question as an intellectual exercise, Luis was increasingly incensed that Bill and other skeptics, including me, did not concede that the impact was the culprit. Frustrated by resistance from much of the paleontological community, Luis concluded that most paleontologists lack proper analytical skills as scientists and characterized the discipline as being analogous to that of "stamp collecting."

As the debate dragged on what struck me most was the moral that it imprinted on my views about how science really works. As a young graduate student with my whole career hanging in the balance, I found myself caught on the side opposite a powerful Nobel laureate. Numerous times during the debate, especially as the impact hypothesis gained credibility within the scientific community, I could see my career being flushed down the toilet, as Bill and I continued to play the role of skeptics. Who would hire a reactionary like me, and even if they did, did I want to spend the rest of my life embroiled in such personally wrenching scientific debates? It became abundantly clear to me that the conduct of cutting-edge science involves the exercise of political power as much as it does scientific testing. This was very disillusioning and took a tremendous emotional toll on me because, by nature, I'm not a confrontational person.

Consequently, I had to find a way to navigate through this highly polemic dilemma by compiling a dissertation that reflected my own convictions without destroying my career, in which I had already invested fifteen years of field work and university training. My salvation came from a former professor, Pete Sadler, with whom I worked at UC Riverside. Pete had developed a statistical model to estimate how often rock layers are preserved in different geologic environments, such as rivers, lakes, shorelines, and deep marine basins.[3] I had taken this model and applied it to various rate-related problems in paleontology. One day during our seminars with the Alvarezes it occurred to me that I could take Pete's model and apply it again to an important issue involving the impact hypothesis.[4] The Alvarez scenario predicted that all the dinosaurs would have gone extinct within one hundred years of the impact, as the result of its deleterious environmental consequences. With Pete's model I could look at the key sequences of rocks containing the fossils that document the faunal changes at the Cretaceous–Tertiary boundary. I could then estimate whether rocks and fossils can be expected to be preserved every hundred years or not. Such resolution is required if these rocks and fossils are to be used to test the impact hypothesis, because rocks and fossils would have had to be preserved every century if we want to truly establish the effects of the impact.

In essence, the rock and fossil records for the history of Earth and the evolution of life are like movies. Pete's method allowed me to look at the layers of rock preserved in Montana and other sites around the world as if each layer was a frame of film in these movies. Since the time of Darwin geologists and paleontologists have realized that these movies are not complete. In any given location on the earth, many frames and scenes are missing from the rock and fossil records, because sometimes rocks are not being deposited at that place, or because rocks that were deposited there have subsequently been eroded. The question is how many frames and scenes are missing from the movie recording what happened during the extinction event.

The results were sobering. At Hell Creek my analysis suggested that we could only expect about one out of every two hundred centuries to be represented by rock layers during the Cretaceous–Tertiary transition at the end of the Age of Dinosaurs and the beginning of the Age of Mammals. The most complete sequence of rock layers deposited on the continents at that time seemed to be in New Mexico, but even there only about one out of every seventy centuries is probably represented by rocks and fossils. This does not mean that an impact could not have caused the extinction. It means that, given our limited ability to tell time so long ago, we probably cannot distinguish between the effects of a sudden impact that killed within a century and the effects of a longer series of major volcanic eruptions that killed over tens or hundreds of thousands of years. So we can believe what we want to believe, but we cannot rigorously test these hypotheses about dinosaur extinction to establish that they occurred as fast as the impact hypothesis predicts.

Fortunately, this conclusion was supported by Bill without unduly upsetting the Alvarez team. I was able to finish my dissertation and graduate at the end of 1983. But the incineration of dinosaurs at Hell Creek had temporarily singed my desire for a career in scientific research.

My own involvement in the extinction debate dragged out for eighteen years. It ended when I published my own perspective of the details underlying the scientific arguments supporting the competing hypotheses for dinosaur extinction in *The Mistaken Extinction*, a book I wrote with Tim Rowe. In essence, my argument was again based on timing. Although the impact theory has gained wide recognition in the public's perception of the extinction and a majority of scientists now accept that an extraterrestrial impact played a role, a significant minority still doubts that the impact caused the extinction. Proponents of the volcanic alternative remain convinced that massive eruptions were responsible.

The "smoking gun"—proving that an impact did indeed occur at the time large dinosaurs disappeared—was found when the Chixulub Crater was discovered buried deep beneath the Earth's surface straddling the shoreline along the northeast coast of Yucatan. Rocks melted in the impact crater date the event to 64.98 million years ago, plus or minus fifty thousand years, essentially the same time that large dinosaurs died out. However, the initial stages of the volcanic activity associated with the Deccan Traps in India began about 68.57 million years ago, plus or minus eighty thousand years, and the last phases of that activity ended about 64.96 million years ago, plus or minus one hundred ten thousand years. Thus, the eruptions and impact overlap, and because many of the "killing mechanisms" associated with the two events were essentially the same, it is very difficult, if not impossible, to clearly decide which was the "true culprit." Some scientists contend that the two events combined in a coincidental one-two punch, and I suspect that represents the most reasonable perspective.

This might sound like a cop-out, but my preference for this explanation is bolstered by facts related to the largest mass extinction in Earth's history. It occurred about 250 million years ago, at the end of the Permian Period, when more than 90 percent of all species were wiped out. Interestingly, this extinction coincided with the largest volcanic eruptions ever inflicted on the continents. This enormous event spewed about 720,000 cubic miles of lava across Siberia, enough to cover the entire planet's surface under a ten-foot-thick layer.

Radioisotopic dates establish that this volcanism began about 253 million years ago, but the main pulse of this activity began about 250.0 million years ago, plus or minus 300,000 years. The time of the extinction was 250.0 million years ago, plus or minus 200,000 years. Thus, within our ability to tell time at the end of the Paleozoic Era, the dates are indistinguishable, strongly suggesting that the eruptions played a role in the extinction. And so far there has been scant evidence to suggest that an extraterrestrial impact occurred at this time. So, to me, to ignore the effects of large continental volcanic events seems a bit foolish.

Even Walter Alvarez, whom I still consider a friend, despite the fact that we argue on different sides of this debate, seems to have similar doubts. In 1997 he wrote in his book entitled *T. rex and the Crater of Doom*:

> Impact as a geologic process . . . must be recognized as a rare but significant kind of event, and evidently the cause of the K-T [cretaceous-Tertiary] mass extinction . . . Can volcanism be dismissed from the list of catastrophic events with global effects? Not yet . . . I would have dismissed the apparent age match between the Deccan Traps and the K-T impact-extinction event as a strange coincidence, if it were not that a second such coincidence has turned up. . . . Recently, Paul Renne . . . has obtained reliable dates on both the Siberian Traps and the Permian-Triassic boundary. . . . they are indistinguish-

able. A good detective shouldn't ignore even a single coincidence like the K-T-Deccan match in timing, and when it is bolstered by a second coincidence . . . it just has to be significant. . . . Right now . . . there is an intriguing mystery, some obviously significant clues, and nobody has any idea what the explanation will be.[5]

Overall, that summary seems like a fair assessment, essentially a plea for a truce while further research is done, and I wholeheartedly support that approach.

The Iridium War has cooled down and my stint on the front lines has ended. I can't say that I conquered, but neither did I crumble. Healed wounds simply show I survived. And for me, it was time to move on.

9

ECHOES OF THE
WILD WEST

By the 1990s my trips to Jordan and the area around Hell
Creek became few and far between, but I occasionally returned
to continue our research. Having finished my dissertation and
worked on an exhibition about evolutionary history at a
museum in San Francisco for three years, my address drastically changed. My job in 1996 involved renovating the halls of
fossils at the American Museum of Natural History in New
York, with some summer field seasons in the Gobi Desert of
Mongolia mixed in. The exhibit renovation turned out to be a
ten-year, $50-million project. Since I was thoroughly enmeshed
in a series of seamless meetings with my exhibition staff, curators, exhibit designers, administrators, architects, and construction managers, Hell Creek and its surroundings were far
from foremost on my mind.

So imagine my dumbfounded disbelief when I wearily woke
up one morning, stumbled to the door, and in picking up the

paper saw a photo of one of my friends from Jordan plastered across the front page of *The New York Times*.[1] John FitzGerald, the only pharmacist in Garfield County, was standing alongside the soda fountain in his drug store where he had made me numerous malts and milk shakes during the searing summers of my student years. Unlike his soda fountain, which dates from 1927, or his Coke dispenser, which hails from 1955, John had unceremoniously come face to face with the advanced technologies of our modern-day media.

As I brewed up some coffee to jump-start my mind, I couldn't help but muse, "What the hell is going on?" A few moments perusal provided some key clues. A knotty contrivance of the Fates had landed John's face on my doorstep.

The situation involved a tense standoff with authorities near Jordan, but John was only tangentially entangled. A nucleus of twenty highly incensed and well-armed neighbors was holed up on a 960-acre wheat farm outside of town. They christened their hideout "Justus Township" and declared that it represented an independent foreign country within the U.S. Somewhat ironically self-designated as "Freemen," they were under siege by a force of more than one hundred FBI agents and other local officers.

Although this standoff brought the whole situation to a head, the tension that led to the siege had been building for years. The roots of the cause, not surprisingly, were money, religion, and politics. In essence, the Freemen were angry at the government for what they felt were unwarranted foreclosures on property long held by their families, and angry, too, about insufficient economic assistance. Their credo also contained race-based and religious-fundamentalist overtones. Recounting the origins of the Freemen's opposition, Carol M. Ostrom and Barbara A. Serrano report[2] that Garfield County's sheriff, Charles Phipps, became suspicious that something was amiss in the spring of 1993. When he drove out from Jordan and delivered foreclosure documents to two ranchers he had known well for years, they didn't seem to bat an eye. They told Phipps that they had just spoken with "one of the smartest folks they'd ever met, and now they had the answers" to all their problems.

It turned out that their guru was the leader of We the People, a group based in Colorado, which advocated that citizens

John FitzGerald in his drugstore in Jordan, Montana. (AP/WIDE WORLD PHOTOS)

in the United States did not have to pay their debts, because the country's currency was "worthless." Paradoxically, he claimed that if the ranchers paid him a fee of three hundred dollars he would help them out of their financial predicament. Essentially, he then proceeded to show the ranchers how "they could pay off farm-loan debt by printing certified money orders on their computers." Phipps was skeptical, to say the least, "I made an excuse to get out of there right away."

The situation that these financially strapped ranchers faced was not uncommon. For decades the owners of many small family farms and ranches depended on federal loans and subsides to make ends meets. But in the late 70s and early 80s, when crop and livestock prices plummeted at the same time that interest rates soared, many owners lost considerable amounts of income and desperately labored to pay off their loans. The government became the enemy rather than the benefactor, and groups such as the Freemen sought to retaliate by filing dubious financial claims against government officials and judges:

> Arguing that the officials had failed to meet their oath of office, the Freemen claimed they were owed money, since the original dollars they had borrowed from banks and the federal government were worthless.[3]

One of the officials that the Freemen around Jordan set their sights on was the county attorney, Nick Murnion. They demanded that Murnion pay them $500 million in minted silver, and he would soon come to realize that the Freemen's demand represented a form of "paper terrorism." But that was just the beginning.

On January 26, 1994, thirty-six Freemen marched into the county courthouse past Phipps and attempted to set up their own government. After the county commission passed a resolution banning the use of the courthouse by the Freemen, they responded by placing a $1-million bounty on Phipps, Murnion, the district judge and three bank officials, threatening to try them and hang them from the city's bridge. Although at first Phipps and Murnion had trouble believing their old friends were serious, the bounty on their lives pushed them well past their point of tolerance, and Murnion filed charges against fifteen Freemen for impersonating public officials. As Ostrom and Serrano summarize, both Phipps and Murnion had family ties around Jordan that stretched back for generations, so they were intent on protecting all the citizens in the region. As Murnion stated, "In a democracy, who's the government? We are, so when you attack the government, you're attacking the people." Despite the fact that numerous locals felt some sympathy for the Freemen's goal to counteract the government's power, most also felt that the Freemen's strategy was misguided:

"I don't know what the right way is, but this ain't gonna do it," says H. K. Riley, 81, whose father homesteaded the land near the spread he still owns. 'They had some good ideas, but they went bad.' "[4]

One of the county's commissioners, Phil Hill, sympathized with the Freemen's plight; however, he had difficulty understanding why the Freemen did not simply react in the same way many other financially troubled ranchers had. Hill confided, "A lot of us got into problems, but most of us just buckled down and dug ourselves out."[5]

Even early on in the crisis, townspeople lamented the schisms created within numerous long-standing and close-knit families. Typical was the plight of rancher and former state senator Cecil Weeding, whose brothers-in-law became enmeshed in the Freemen's activities. As he told a reporter, "They thought we should have joined up with them. They're basic down-to-earth people, but they just got into this cult. You can't reason with them."[6]

This sentiment, expressing regret as families watched relationships with their next of kin torn asunder, became commonplace in news reports over the next few years, as well as more angry sentiments demanding armed action to roust out the Freemen. The potential was clearly there for tragedy on a scale equal to that conceived by the greatest Greek playwrights. Some people were acting quite irrationally, at least in terms of most conventional social standards, and it was disturbing to see the fabric binding the society together in this rugged outback rapidly unravel. The prospect of bloodshed and multiple fatalities hung grimly over the landscape, fueled not only by the immediate death threats but also by the specter of recent, violently terminated standoffs at Ruby Ridge and Waco.

As summarized in an article in the *Billings Gazette* on March 18, 1996, many of the townspeople grew frustrated waiting for

Vehicle on grounds of the Freemen compound northwest of Jordan, Montana, the day
the siege began on March 25, 1996. (COURTESY OF THE *BILLINGS GAZETTE*)

the federal authorities to intervene, a circumstance that left
many feeling both angry and helpless. "Given the option to
hang them or send them to trial," said Phipps, "I think most
people here would hang them."

Yet, the town's frustration was tempered by honest concerns
about how it might end. With so many families straddling the
growing chasm between the Freemen and the authorities, the
community's sense of painful foreboding was acute:

"There will probably be bloodshed before all of this is
finished," said a patron at the Hell Creek Bar one recent
night, as others around him nodded in agreement.

"I hate to see it happen, but I'm afraid it's going to have to," said another. "And when somebody gets shot out there, it's going to be somebody we all know."[7]

The standoff near Jordan began in earnest on March 25, 1996, after two Freemen leaders were arrested when they strayed off the confines of an associate's ranch where they had been holed up for over a year. As many as twenty Freemen and their family members were sequestered, including three young girls aged eight to fourteen.

According to accounts provided by locals and negotiators, the Freemen were not only heavily armed but well supplied. They had several months to prepare for the siege, and estimates for how long the Freemen could hold out ranged from three months to years. In addition to whatever store-bought food they had stocked, the farm also contained livestock and fishing ponds, as well as $80,000 worth of harvested wheat. Moreover, the Freemen had generators, cellular phones, radio communications equipment, and satellite television reception.

Most people were pleased that the FBI had finally acted. One soapbox for public reaction proved to be the counter of John FitzGerald's soda fountain. Initially the tone was optimistic, as townspeople gratefully thanked Murnion for his steadfast determination to resolve the situation. Meanwhile, a whirlwind of government agents invaded the small community, prompting one resident to quip, "There hasn't been this many of them in Montana since Custer."[8]

The Freemen compound where the standoff played out is only thirty miles northwest of Jordan—essentially next door to where Bill and our Berkeley crews worked. By "Big Sky" standards, it's a stone's throw from the McKeever ranch, where we spent weeks collecting exquisite fossils of dinosaurs and diminutive mammals along the steep ridges next to Billy Creek. The Watts ranch, where Harley collected a fragmentary tyrannosaur jaw, is just to the east. The Hauso and Murnion spreads, where we first located the iridium-rich impact layer, are essentially just over the nearby buttes. So I personally know many of the people whose normally serene and isolated existence was violated in the vortex of that invasion.

Although many Jordanians were happy to see actions being taken, sentiment regarding the Freemen led to disagreements among those outside the Freemen sphere:

> "Some people are mad because they don't think [merchants] should deal with them," said Stanton, the county clerk. "But, geez, somebody walks in with cash and wants to buy your groceries—what are you going to do?"[9]

Beyond that, there were still members of the community that supported the Freemen to some degree:

> "Quite a few of us in the county don't mind them being here," said one woman who lives near Justus Township. "'Some of us sympathize. We understand where they're

coming from, but we don't understand what they're doing."[10]

Meanwhile, most residents attempted to get on with their never-ending work. Fields still needed to be plowed and livestock still needed to be fed. Rancher Dean Rogge summed up the prevailing opinion:

"I would say about one percent of the people support them," Rogge said. "In a county of this size, one percent doesn't amount to a lot of people."[11]

Rogge went on to note that the people of the county had long been known for their industrious work and community spirit, guidelines that the Freemen violated through evading taxes after accepting large government subsidies:

"Everyone thinks taxes are too high," he said as he drove his beat-up Chevy pickup around his property. "But we're law abiding citizens, and we pay them."[12]

From the start of the standoff, national and local media engaged in a feeding frenzy. Reports indicated that at one point the number of reporters and federal agents on the scene exceeded the total population of Jordan itself, which at the time was usually listed between four and five hundred. ABC alone sent twenty-six employees in anticipation of a catastrophic climax.

No one who ventured outside was immune from the media's onslaught, and even seeking refuge inside a public establishment was pretty much pointless. Nick Murnion clearly bore the brunt of the media's barrage. As of April 12,[13] his scorecard of interviews was dwindling down to a couple per day, in contrast to the last week in March, when he logged numerous appearances on national and international radio and television shows. "Day four—March 28—is still kind of a blur," he said. "It started at dawn with an interview on *Good Morning America* and extended after nine at night with a conversation on San Francisco talk radio. The day also included snippets with Australian and German television, among others."

The favored hangout for all these foreigners was, not surprisingly, the bar. For years after the siege Jordan's Hell Creek Bar, which we frequented as students during out field seasons, and where Harley successfully sleuthed for clues to discover his tyrannosaur, was rimmed around the inside by caps donated by the agents and correspondents involved in the siege.

The proprietor of the Hell Creek is Joe Herbold, an amiable, middle-aged master of ceremonies who manages to move effortlessly between his indigenous patrons and guzzling guests. By nature he's giftedly gregarious, and he enthusiastically assumed the role of trying to keep everyone's spirits up. As de facto town greeter, he was unfazed when *Good Morning America* filmed inside the Hell Creek. Instead, he focused on a bevy of correspondents just entering the bar, exclaiming, "We want to know who you are and where you're from. Get over here and sign in." Joe had started a pool in which people registered guesses as to how many journalists would show up. His own guess was seventy-five. In the midst of the chaos Joe saw Jordan's glass as half full rather than half empty. "It gives us a chance to show off our little community. Under different circumstances, this would be more fun than you could shake a stick at." Whereas, Murnion simply sensed the absurdity, "I don't think we've got enough people in Jordan for everyone to interview," he added laughing.[14]

Speaking more directly from the confines of John FitzGerald's drug store, Sharon Nelson, John's longtime assistant, responded to the media onslaught with a concise statement laced with a dose of anxiety. "We're ready for them to leave and for things to get back to normal."[15]

But it would be a while before that happened, and in the meantime many merchants embraced this capitalistic opportunity. For, on the other hand the media influx generated a welcome economic boom for the local motels, restaurant, bars, and shops. This led to some perplexing reactions, as the community came to know and even enjoy the company of some of their new clients. As one teenager told CNN, "I wish they could have stayed, 'cause I could have worked longer and made more money, and it's better business for the town."[16]

But underneath all the hoopla the newspaper accounts of the day recorded the misgivings of townspeople about the images of their city that this ordeal projected to the world. All but a handful of these people are decent, law-abiding citizens who simply wanted to be left alone, and the spectacle of the standoff with the radical rebels in their midst understandably disturbed them.

The population was used to taking care of its own problems without the help of federal agents. It's become a time-honored tradition in a remote landscape, which for more than a century has harbored the likes of "stagecoach robbers, renegade Indians fleeing reservation life, and even Butch Cassidy and the Sundance Kid, who hid in the gullies that crease its surface."[17] The vigilante tradition still runs deep, and the community was most perturbed by the way the Freemen mistreated their neighbors by foisting bad checks on them. As Dale Fellman put it:

"In the old days, 'If you saw a horse thief talking to someone you just assumed they were both horse thieves and hung them both. And if the FBI hadn't got here now, the people would have taken care of it themselves.' "[18]

As the siege dragged on, trying the patience both of citizens and federal agents, a totally unrelated and unexpected occurrence turned the attention of both the media and the minions of law enforcement to another locale in Montana. On April 3 a suspect alleged to be the Unabomber was arrested. Media crews already positioned in Jordan scrambled to cover the breaking story. Jordan's usually sleepy airport hummed as crews from the national networks chartered planes and flew out to the small town of Lincoln in the western part of the state. " 'Nothing happens in Montana,' wisecracked a weary looking ABC staffer."[19]

The simultaneous realization that two entities of outlaws were operating in the state threw many residents of Montana into a tizzy. Another article in the *Billings Gazette* on the same day attempted to sum up the sentiment, while the populace persevered by falling back on their highly honed sense of humor. Previous nicknames for the state, such as "The Last Best Place," "The Big Sky Country," and the "The Treasure State," momentarily gave way to more cynical ones, such as "Big Scare Country," "High, Wide and Wacky," or "Montana: An Extreme State of Mind."[20] Newspapers and travel agencies fretted about whether the state would lose its romantic luster.

Some folks in Jordan joined in by acknowledging the seeming absurdity of the situation. One byproduct of the siege was that some of the dirt roads of the region were doused by road crews with a welcome veneer of gravel. I can attest to the fact that when it infrequently rains at Hell Creek the roads are transformed into impassible quagmires of gumbo—extremely slippery mud. As my paleontological colleague, Mike Greenwald, once warned me as we walked out into the outcrops after a rainy night, "Be careful; when it rains the area turns right back into a Cretaceous swamp." So graveled roads are a true luxury. At least one FBI agent was killed trying to negotiate an ungraveled track, and the locals approved of the logistical ungrade, "Garfield County residents need a few more standoffs," one man joked, "so they can get some more roads graveled."[21] T-shirts in town also reflected the mood. The backside of the latest one featured the word HATE, which had been crossed out, followed by the phrase NOT IN GARFIELD COUNTY. The front sported the question, HAVE YOU HAD YOUR INTERVIEW YET?

April morphed into May without the standoff being resolved. On May 29, shortly after the siege passed the two-month mark, the *Billings Gazette* published a poll regarding the events in Jordan and its effect on Montana's image. Eight hundred citizens, equally divided between men and women, were interviewed in various regions of the state. The first question was: Do you approve or disapprove of federal and state law enforcement officials' decision not to arrest the Freemen in northeastern Montana? Thirty-two percent approved, 51 percent disapproved, and 17 percent were unsure. The second question read: Generally speaking, do you think Montana's image has been tarnished by Freemen, militia, suspected Unabomber, and other antigovernment groups? Thirty-four percent said yes, 55 percent said no, and 11 percent were unsure.

In the meantime some citizens of Jordan took it upon themselves to spur a resolution by circulating and signing a petition demanding that the government take action. Their patience was wearing thin, and the document advocated the use of "reasonable force" before a deadline to be determined in order "to ensure" a surrender. Two hundred citizens signed the petition. One leader of the petition drive, Brent McRae, indicated that "People still feel the threats of intimidation but most are saying 'intimidation be damned'—they are tired of it."[22] He noted that things had gotten so bad that, "There were girls' basketball teams who called to see if it was safe to come to Jordan. It got that spooky for some people."

The citizens clearly wanted action before irreparable damage was done, and on June 3, with the backing of the local community and the patience of law enforcement continuing to wane, climatic events began to unfold. The electrical power to Justus Township was severed, while generators continued to feed the surrounding ranches. On June 4, the seventy-second day of the siege, the standoff with the Freemen eclipsed the longest previous siege in the nation's history. Ironically, in relation to Hell Creek's earlier history, that record had previously been set by militant Native American descendants of Sitting Bull at Wounded Knee in South Dakota over seventy-one days in 1973. By June sixth, two parents emerged with their eight- and ten-year-old daughters and left the compound with an escort of federal agents.

A week later, on June 13, after torturous weeks of negotiation, the remaining Freemen on the isolated ranch surrendered in pairs to federal agents, "in a cordial, orderly process that took place over a cattle guard and began with Freemen leader Emmit Clark shaking hands with government representatives."[23] The siege had lasted for eighty-one days, and now the healing process began.

The weary relatives and residents expressed differing sentiments concerning the potential for reconciliation. Some were more optimistic than others. When the standoff started, Dean Rogge, as well as former state legislator and rancher Cecil

Weeding, had expressed pessimism.[24] Rogge stated, "We've always been together and helped each other out . . . Now we live looking over our shoulders. The community will never be the same." Weeding expressed similar feelings, having suffered a schism in his own family that "has torn the community apart. It will never be the same. There's just too many ill feelings. Those things will not heal in our lifetime, I'm sure."

More hopeful was Carol Hellyer, the dispatcher in Garfield County's sheriff's office whose sister, Agnes Stanton, sought refuge inside Justus Township. She had stated on April 7 that she looked forward to eventually welcoming her sister back into the family. "We're a pretty forgiving community. We take care of our own. Aggie has not done anything that cannot be repaired."[25]

Probably the most poignant rupture among relatives was, indeed, between the son, father, and grandfather of the Clark family. Even in the aftermath of the surrender, reconciliation remained elusive. Well before the standoff began, Dean Clark had purchased land eventually included in Justus Township, when authorities had foreclosed on the property. It had been owned by his father, Richard, and his grandfather, Emmit. But, when the siege began the father and grandfather kicked the son, who was not a Freeman, off the land and threatened to arrest him for trespassing if he returned. During the standoff the son had no access to the grain he had stored on the property the previous year, and he was unable to prepare the fields for the spring planting. His financial situation became quite dire as the siege dragged on. Not until after it ended did he gain sufficient access to plant his crop and sell his grain, but his attitude toward his father and grandfather remained skeptical, as he stated while he plowed his fields on June 15:

"I don't know. . . . He's done some pretty tough things to me. I guess there's always a slim hope, but it's been his choice all the way.

"I'm afraid it's probably a long way from over," he added. "Hopefully they'll quit torturing me."[26]

Nick Murnion was especially thankful that it had all ended, although he felt that the federal agencies should have acted more quickly. On June 14 he said that the end of the siege "means life getting back to normal. It means after two years and two months these people will finally be brought to justice." Brent McRae observed that the crisis "has overtaken the county and its residents. For the most part, the majority of the people and businesses were ready to see this thing over." Murnion concurred, "I think we all wanted to see it end earlier." McRae agreed, but emphasized the significance of a nonviolent conclusion. "None of these people were deserving of martyrdom. I think the people here are relieved that no one did lose their life."[27]

With the standoff concluded peacefully, the townspeople breathed a sigh of relief and rejoiced in a traditional way that I had become quite familiar with as a student in the early 1980s. On June 15 they blocked off Main Street and held a public dance in honor of a reunion for former high school students. Many citizens expressed their joy about the return to normalcy.[28] "I'm glad it's over. I'm glad it ended peacefully, and not like Waco," said one. Yet, some residents continued to express differing opinions about the consequences that the Freemen should face. For example, one said, "These are our ranchers, our friends that we have known for forty years." But another was not so accommodating, "I hope they get some of the punishment they deserve. But thank goodness it's over." Yet, another yearned for financial normalcy to quickly return. "Whenever this thing started, a twenty-ounce pop cost sixty-five cents, it's gone up to seventy-five, and it's like we must be in downtown Billings somewhere." With its days in the national and international limelight ended, it seemed inevitable that Jordan would once again morph back into a small rural community deep in the outback of Montana.

Did it? On the five-year anniversary of the standoff's conclusion, the *Associated Press* delved into the aftermath. Al Bassett, the new owner of the land and buildings where the Freemen holed up, still recalled that "It was like a civil war." Some townspeople, including Claudia Stanton, had indeed reconciled with the former renegades. "You can't harbor hard feelings, or there's never unity," she said. "We're forgiving people, or we wouldn't live on this land." Yet, others were not sure that the Freemen had, in turn, forgiven their former opponents in the community. Julie Loomis lamented, "I think people have forgiven them. I don't think the Freemen have forgiven us for not joining them."[29]

Few vestiges are still on view from the days of the siege. When I returned to Jordan in May 2003 to conduct research for this book, Joe Herbold had long since conceded his bet concerning how many media would show up for the siege. He confirmed to me what had previously been reported; he had drastically lowballed his guess:

> The tally sheet at the Hell Creek Bar . . . cracked the two hundred mark Tuesday night, prompting owner Joe Herbold to buy the fellow from CNN in Atlanta a beer on the house. There were cheers all around.[30]

Joe had also taken down the caps from around the Hell Creek Bar. He really didn't need them anyway, because the bar sells great caps of its own. Stitched to the front above the bill is the bar's logo, a vertically ascending wall of red and orange flames encircling the name of the bar and its location. In the aftermath of the siege, I bought one for a friend who lives in New York, who wore it to work one day on the subway. As she sat reading, a teenager sitting across from her stared in rapt admiration and asked her where she got such a cool hat. Thinking he'd heard of the recent turmoil in Jordan, she asked if he'd been keeping up on the Freemen siege. "Freemen siege?" he replied with wonderment. "No. Those are my two favorite athletes . . . Jordan and Montana."

So, although the events in Jordan were traumatic, many outsiders did quickly lose interest, which seemed just fine with the local citizens. Few of my old friends seem interested in discussing their memories.

When I asked Jane Engdahl about her impressions of the siege, she shrugged her shoulders and said she felt fortunate that it didn't affect her family more directly. Although the Engdahls live within several miles of the former Freemen compound, she said they didn't see many FBI agents or their vehicles pass by their place. Similarly, when I spoke with Judd and Jay Twitchell, they said that they're just glad it's over and that things have returned to normal. Jay pointed out that one of their neighbors, who was not involved in the siege, but was charged with another related offense before the standoff, had already served his time and returned to life on his ranch. When I stopped in at John FitzGerald's drug store for an ice-cream soda, I kidded him about the day I picked up the *New York Times* to see his face looking right back at me. He simply nodded and responded, "That was a bad time." In all, my old friends much preferred to catch up with my exploits and discuss the problems that continue to bedevil life at Hell Creek.

Throughout all the commotion, the townspeople never lost sight of their basic concern about life in Jordan. As Timothy Egan wrote in the *New York Times*,

> . . . all around the island of the siege, people are engaged in a daily struggle to keep their little hold of fast-fading American life from disappearing from the map altogether. Real drama, says rancher Bev Murnion, her face

turned out of the lacerating wind, is getting $300 for the same size cow that brought $500 two years ago, or watching a newborn calf die after a sudden cold snap. A real siege is when a chinook wind melts a foot of snow, and then the standing water freezes at night, and Ruth Colter has to drive 30 miles to deliver the mail to three patrons— down from more than 50 mail stops when she started the job 47 years ago.[31]

Life in modern Jordan is indeed tenuous and remote. That fact was reinforced on the morning that I picked up the newspaper to read John FitzGerald telling Timothy Egan:

If this business folded, I don't know that Jordan would last much longer.... You see here, from the Missouri River to the Yellowstone, from Glendive west for 240 miles, I'm the only pharmacist.[32]

These reflections on modern life at Hell Creek, compared to when Lewis and Clark first struggled through the area, lead to some interesting, if perplexing, visions of what lies ahead.

10

IS THE PAST THE KEY
TO THE FUTURE?

Feeling privileged to have learned about and even to have participated in some of the remarkable events surrounding the history of Hell Creek, such as the debate about dinosaur extinction, I'm left to wonder what's in store for the future of this legendary region. After the invasion of the modern-day marshals and the media frenzy during the Freemen standoff, will it inevitably be forced to march in lock step with the rest of us into the technological fantasyland of the twenty-first century?

On the one hand, there are some indications that it will. As Jordan's residents wrestle with the pressures threatening to push them out of their tenuous foothold in this unforgiving landscape, technological changes are playing a pivotal role. Those who seem best suited to survive are attempting to weave together elements of their rural past with technological trends leading into the future.

As a result of the Freemen standoff, County Attorney Nick

Murnion is one citizen of Jordan who has found his feet firmly planted in both worlds, somewhat as I did when I first stood on the precipice of the badlands gazing down on the past. But his perspective is just the opposite. He seems to be seeking out an oracle to gain insight into the future, not the past. He has seen the luminosity of the light that modern society can cast on its citizens with laserlike intensity. In the aftermath of the standoff during April 1998,[1] a woman from Boston named Caroline Kennedy called Murnion at this office. As Murnion remembers it, "At first, I didn't key in on the name. Then she started talking about this Profiles in Courage Award." He eventually realized he had won a prestigious award, and once again asked the woman who she was. "Caroline Kennedy. The president's daughter."

Murnion had indeed won "something really spectacular," the 1998 Profiles in Courage Award, which was established by the John F. Kennedy Library Foundation in 1989. It goes to an elected government official "who has made decisions based on principle, despite resistance from local constituents, special interest groups, or adversaries."

Despite the honors and horrific threats associated with the siege, however, Murnion remains focused on the future. In addition to serving as county attorney, Murnion is also a former president of the Mid-Rivers Telephone Cooperative, which serves Jordan and its neighboring cities. He is dedicated to maintaining the viability of his community as it enters the new millennium. As he indicated to Timothy Egan,[2] "If Garfield County can ever shake the spotlight brought by the Freeman standoff, it may have a future in the new-technology economy." Like officials in numerous other small towns in remote corners of the United States, Murnion, who then served as a trustee on the board of the local telephone co-op, was trying to recruit high-tech businesses to set up shop in Jordan. The community's technological infrastructure is indeed expanding, as Egan notes:

> A side benefit of having the hordes of FBI agents around," Mr. Murnion said, "is that they brought in a cell that made cellular phone calls possible. Jordan is now connected to the world.[3]

This may seem intensely incongruent, in light of Jordan's history and agricultural economy. But even some of the ranchers have gone high-tech, sowing and reaping their wheat from the air-conditioned comfort of their tractors, equipped with satellite-based gadgetry designed to maximize their yields. Yet, they are members of an "endangered species." Murnion knows that

> we desperately need people. Most of the farms used to have a husband, wife, a son or two, and a ranch hand, but now they're all getting down to just one or two people. It takes twice as much land and twice as many cows to make the same living you did ten years ago.[4]

Indeed, during my pilgrimage to Jordan in May of 2003, Bob Engdahl told me that if a ranch doesn't incorporate ten thousand acres the owner can't really make a go of it. He now works seventeen thousand, and even that amount is sometimes perched perilously close to the cutoff point—for, others feel that twenty thousand acres are required for a viable spread.

So Jordan's fate hangs by some ominously thin threads, and some intriguing signals involving the cultural and social trends within the northern Great Plains suggest that Hell Creek may not leap effortlessly along with most of the world into the touted "Age of Technology." In contrast to the geological principle introduced earlier, which states that "the present is the key to the past," some have argued that Hell Creek's past may be the key to its future. To resolve the focus on this issue and this temporal paradox, it is necessary to provide some further historical perspective.

As Don Baker wrote in *Next Year Country*, census figures showed that Montana hosted a population of about twenty thousand in 1870, including fewer than one hundred Caucasian women in the eastern part of the state near the Hell Creek region. By 1880, after Custer and Sitting Bull battled it out at the Little Bighorn, the state's population had almost doubled to almost forty thousand; by 1890 it more than tripled to over one hundred and forty thousand. In 1900, a year after Arthur Jordan established his town just south of Hell Creek, Montana's population soared to nearly a quarter of a million. By 1910 it expanded to more than three hundred and seventy-five thousand, and the stage was set for the population boom of the teens.

Throughout all of these decades the precipitous rise in population rode on the back, in large part, of European immigrants moving westward from the eastern regions of the United States. Jordan himself was a perfect representative of this trend. In eastern Montana, Fins, Poles, Danes, Germans, Ukrainians, and Scandinavians all established ethnically based communities. Even Asians who helped build the transcontinental railways eventually settled in the state upon the project's completion. The Northern Pacific, Great Northern, and finally the Chicago, Milwaukee, and Puget Sound lines all crossed eastern Montana by 1909, and each enthusiastically trumpeted the glories of the land newly opened up for settlement. As Baker recounts,[5] embryonic communities along the railroad lines created colorful murals and billboards near their stations in order to advertise the assets of the area and attract potential settlers passing through. They also established commercial clubs, precursors to later chambers of commerce, to promote local agricultural and other possible careers. Once again, tribes of Native Americans paid a price in this expansion by having their reservations downsized to make room for the new settlers. Land agents, often employed by the railroad, met trains filled exclusively with potential immigrants seeking cheap or free land. The agents were armed with promotional materials and maps of the region describing what parcels were still up for grabs. The agents then whisked the new arrivals out to the chosen parcel, which sometimes bore scant resemblance to the utopian landscape that the agent had described.

Government legislation fueled this land rush. As early as

1841 an act called the Preemption Law allowed settlers in Montana to claim one hundred and sixty acres of unsurveyed land for a dollar twenty-five per acre. The initial filing fee was twenty-five cents per acre, and the remaining dollar per acre was due at the end of eighteen months. The Homestead Act of 1862 provided even more generous terms. Any adult who had not fought against the United States could, for a twenty-five dollars filing fee, claim one hundred and sixty acres of public land, providing he would live on the land for five years, farm on some of it, and make a few improvements. This act had minimal effect on Montana, however, because more fertile and more easily accessible land was still available to the east, where attacks by Native Americans were less likely. It wasn't until fifteen years later, just after the slaughter at the Little Bighorn when Miles was chasing Sitting Bull through the Hell Creek region, that the government passed the Desert Land Act of 1877, which catalyzed a land rush to Montana. By then it had become clear that one hundred and sixty acres of dry land was insufficient to support a self-sustaining farm or ranch on the desertic prairie. So this legislation made it possible for a person to claim six hundred and forty acres of nonirrigated land for a dollar twenty-five an acre, if he would irrigate the land and start cultivating it within a three-year period. Although this drew many people to some parts of Montana near water (more than sixty thousand filing claims were made in 1881 alone) much of the land near Hell Creek could not be easily irrigated.

Finally, in 1909, a decade after Jordan founded his town, legislation was enacted that paved the way for a population boom near Hell Creek. As noted by Dayton Duncan in *Miles From Nowhere*,[6] it appeared that the arrival of railway service was imminent when two companies surveyed for routes through the area. In The Enlarged Homestead Act of 1909 Congress expanded the size for claims in the unirrigated lands of the western United States to three hundred and twenty acres. By 1916 the limit for claims was doubled to six hundred and forty acres by the Stock Raising Homestead Act. Profits for farmers flourished during World War I, as demand increased for wheat and patterns of unusually high precipitation made the prairies inordinately productive. As the number of homesteaders also boomed the diversity of wildlife in the region diminished. For example, the last sighting of an Audubon's big horn sheep in the U.S. occurred near Hell Creek in 1916. By 1920 the population of Garfield County swelled to 5,368 in the midst of this boom.

When I was a student at Berkeley both the Engdahls and the Twitchells were kind enough to give our field crew access to two of the cabins that these homesteaders built. The Hawkins cabin, a luxurious two-room affair, now sits about a mile from the Engdahl ranch house. It provided us with much appreciated shelter during the tempests that occasionally thunder across Hell Creek. "Stubby's" cabin on the Twitchell ranch, although comprising only one room, provided similar accommodations. For a pampered city boy like me it was almost impossible to envision the early homesteaders trying to survive the minus-sixty-degree temperatures that many of those early winters brought.

The Hawkins cabin on the Engdahl Ranch north of Jordan, Montana. This cabin served as the base camp for field crews from UC Berkeley from the 1970s through the end of the century. (LOWELL DINGUS)

In July of 1893 the Pulitzer Prize–winning historian Frederick Jackson Turner spoke prophetic words to the American Historical Association. They would change the way Americans thought about the West:

In a recent Bulletin of the Superintendent of the Census for 1890 appear these significant words: 'Up to and including 1880 the country had a frontier of settlement, but at present the unsettled area has been so broken into by isolated bodies of settlement that there can hardly be said to be a frontier line. In the discussion of its extent, its westward movement, etc., it can not, therefore, any longer have a place in the census report.' This brief official statement marks the closing of a great historic movement.[7]

According to Turner the frontier was no more. In 1920 Montana's population totaled almost five hundred and fifty thousand. All of this expansion seemed to fit the model and thesis that the United States had essentially settled the full extent of its frontier. Even though Arthur Jordan had not yet moved to the banks of the Big Dry, Turner would have considered the area around Hell Creek simply an isolated vestige of the once-great and imposing frontier. As Dayton Duncan goes on to recount, the future for Jordan seemed bright.[8] Promotional material published by one railroad, entitled "Garfield County in the Pacific Northwest, Garfield County in the Corn Belt," sought to recruit more immigrants to the region, even though no tracks had been built. As Duncan states, "Jordan billed itself as 'the natural metropolis' of this new Eden, with 'the resources to make it one of the most active and important commercial centers in eastern Montana.'"

There were two basic problems with this pie-in-the-Big-Sky attitude—conniving railroad executives and inconsistent precipitation. The railroads never did build a line through Jordan, despite promises to do so, and the unusually wet years in the teens lapsed into the arid decades associated with the Dust Bowl. As Duncan relates, Arthur Jordan decided that his nascent municipal creation was no longer for him and moved out of the town to the mouth of Hell Creek, where it empties into the Missouri. Eventually, in 1922, his restless soul once again got the best of him and he abandoned the area altogether. There were just too many people to please his love of the open country. As Duncan relates,[9] 1920 represented the zenith of Garfield County's population boom, even though there was still only about one inhabitant per square mile. As more normal and arid climatic conditions returned, lower agricultural profits did as well, leading to an accelerated exodus from the region in the 1920s. Massive desertions followed in the 1930s, as the drought continued and the nation fell into the Great Depression. The population had plummeted to 2,641 by 1940, initiating a trend that would continue throughout the rest of the century. By 1990 this county, about the same size as Connecticut, counted 1,589 inhabitants, or about one person for every three square miles, and the town that Arthur Jordan founded a century before struggled to hold on to its tenuous grip as Garfield County's only city.

As defined by technical terms used in the sciences of geography and sociology, there are two major types of settings in which people live on the Great Plains. John Alwin, in his 1981 article in the *Annals of the Association of American Geographers* entitled *Jordan Country—A Golden Anniversary Look*, describes these as "sutlands" and "yonlands":

Jordan Country clearly remains a yonland, distinctive from adjacent sutland areas. Both terms were introduced by rural sociologist Carl Kraenzel in the 1950s . . . Sutlands are more densely settled areas, often linear in form. They are usually strung out along and adjacent to major lines of transportation and communication, often paralleling major rivers. . . . The yonland is an area of significantly lower population density. . . . Such areas are literally "out yonder." Here dry land farming dominates, where irrigated agriculture predominates within the sutlands.[10]

However, the trend toward greater yonland formation is not, by any means, limited to the area around Hell Creek. In 2001, Timothy Egan revealed:

More than 60 percent of the counties of the Great Plains lost population in the last ten years. An area equal to the size of the original Louisiana Purchase, nearly nine hundred thousand square miles, now has so few people that it meets the Nineteenth century Census Bureau definition of frontier, with six people or fewer per square mile.[11]

Yet, that was not the only notable trend Egan found. Across the Great Plains the land is returning to a state more like before the mid-1800s, including a concomitant increase in the presence of Native Americans. Examining these developments through the long-term lens of history, they suggest that the attempt to turn the prairies into an agricultural Garden of Eden may have not only been somewhat misguided, but also a temporary phase. With the flight of white farmers and ranchers from the region, Native Americans are returning. In fact, they are the only demographic entity to have registered a significant population increase in the region. Although these trends have existed for decades,

> they have reached a point—one hundred and eight years after Frederick Jackson Turner suggested that the American frontier was closed, with the buffalo herds wiped out and native populations down to a few tribes—that there are now more Indians and bison on the plains than at any time since the late 1870s.[11]

Has the seemingly irreversible engine of American progress been rudely thrown into reverse? Will the area around Hell Creek once again return to a more pristine and unpopulated landscape of grassy prairies and barren breaks inhabited by bison and grizzly bears? Some are so bold to foresee that scenario. As Duncan relates,[12] in 1987 a nonprofit organization located in Missoula, called the Institute of the Rockies, suggested that a large swath of eastern Montana, including all of Garfield County, be transformed into the Big Open Great Plains Wildlife Range. Fences would be banished throughout fifteen thousand square miles, as would domestic livestock. Previously natural denizens, including seventy-five thousand buffalo, forty thousand elk, forty thousand antelope, one hundred and fifty thousand deer, along with their natural predators, such as the grizzly, the mountain lion, and the buffalo wolf, would replace them. Instead of heavily subsidized farms and ranches, which the Institute argued had destroyed the environment, the region would be returned to more natural habitats. They foresaw that the wildlife range could serve as "an international destination, a magnet for tourists, sportsmen, photographers, and outdoors enthusiasts" drawn to immense free-roaming herds "unequaled by any other place on earth save possibly the Serengeti Plain of East Africa." The area's economy would benefit from the thousand new tourist-related positions that would be generated, and by transforming a ten-thousand-acre ranch into a hunting range, a landowner might earn almost fifty thousand dollars annually from fees for permits.

It is an audacious idea, but there are several problems, not the least of which was that none of the proponents engaged the ranchers of Hell Creek in the discussions about the project's development. Besides, several of the families that I know, including the Twitchells, and the Trumbos, who own the ranch immediately to the east of the Engdahls, already get fees from hosting sportsmen who wish to hunt on their land. The ranchers are fiercely protective of both their land and lifestyle, and from what I've been told, are not anxious to cede either to another invasive government program. Adding to the humiliation, as Duncan reports, "Jordan learned about the proposal when someone sent a *New York Times* article about it to the president of the local chamber of commerce." This served to reignite all the resentment that the locals hold toward the government and outsiders, who the ranchers felt already exerted an unwelcome level of control over their lives. Enraged by the implication that they did not know how to manage their own land and affairs, tension quickly mounted. "What upset people was that someone from the outside came and said the people had wasted their lives here," remembered Scott Guptill, who runs the local paper with his mother. "People were really insulted."[13]

Although some merchants were receptive, the depth of the general dislike became obvious once the chamber of commerce set up town meetings with the proponents. At one in Brusett, a small outpost about twenty miles northwest of Jordan near where the Freemen's compound was located, a vehicle believed to belong to one of the project's sponsors had its back window shot out. The die was cast, and several county organizations and politicians lined up against the proposal. Yet, it is not entirely clear whether some version of the project will eventually come to pass. As one merchant in Jordan tells Duncan, "Deep down, a lot of people think it's probably inevitable, but they'll fight it to their dying breath."[14]

John Trumbo, on whose land I often wandered as Bill Clemens's student at Berkeley, probably summed up the ranchers' sentiments best when he said:

It's like someone told us here in Garfield County that they were going to take part of New York State, where no one lived, and turn it into a moose preserve. Not knowing anything about the place, it'd probably sound like a good idea. But around here we don't need more wolves, we don't need more coyotes, we don't need a Big Open. Everybody's got an idea for us. We just want to be left alone.[15]

Perhaps John is correct, and he along with the Engdahls, Twitchells, and other undaunted ranchers will tenaciously hold onto their land and lifestyle. They've successfully done it to this point in 2004, but I wonder whether they will actually have the final say.

The thought of the area returning to its former majestic state is truly entrancing. In fact, some of the Hell Creek region is being preserved in relatively pristine condition. The boundary of the Charles M. Russell National Wildlife Refuge surrounds the banks of Fort Peck Lake, which now inundates the bed of the Missouri River where Lewis and Clark explored. Grizzlies no longer maraud through the Breaks, but elk, eagles, and hundreds of other native denizens do. The real question is whether more of this long-lost landscape will return to a more original state.

For a paleontologist interested in the dimly lit dimensions of the deep geologic past, the chance to sense firsthand what it was like, even just a geological instant earlier than the present, would truly be a heavenly experience. Yet, it is beyond me to wish ill on the Engdahls, the Twitchells, and the other rough-hewn ranch families that so graciously took me under their wings and sheltered me from the perils of Hell Creek. Their hospitality, along with the hostile yet radiant beauty of the landscape, has irreversibly altered my views of myself, humanity, and species from the distant past. The branding physically battered my body, and the Iridium War sapped my mind. Yet both are tests that I'll always treasure. Hell Creek has tempered my character, and its people have played a large part.

In doing the research for his essay on the Big Open in 1990, the then-publisher of *Time* magazine, Hugh Sidey, spoke with a rancher named Bill Mathers, whose spread lies to the south outside Miles City:

> Mathers, 66, has seen the land ravaged by the plow, the water sucked from the aquifers and wasted, the oil and mining industries nose-dive, and the children of the plains rush for the rural exits. He was in the Montana legislature for 20 years.
>
> In the end, Mathers believes, land governs almost everything else. "You work with the land," he says. "You can't work against it."[16]

Although I agree, I prefer to state it alternatively. I can't truly foresee what the future holds for Hell Creek. In fact, I am only convinced of one outcome. As they have for billions of years, the majestic breaks and parched prairies, along with all the inhabitants that happen to live there, will continue to change. For, the Earth and all of us on it inexorably evolve.

EPILOGUE

I have an opportunity to put this thesis to the test in May and July of 2003, when I return twice to the prairies and coulees around Jordan in order to round up the final bits of research required to finish this book. Immediately upon my arrival it becomes amusingly obvious that evolution continues to direct the scenes and actors across the vast spectrum of personal and environmental scales involved in events at Hell Creek.

I'm to rendezvous with both Bill and Harley out at the Engdahls', where we used to camp, in order to make the old rounds. Although I still head for the field a few times a year, I've noticed that the ground is not getting much softer, so I've shamelessly booked a room at Fellman's Motel to provide a bit more insulation.

As I check in, the manager casts a discerning eye in my direction and speaks a phrase I've not heard in two decades, "You one of the bone-diggers?"

Instantly busted, I confess to my fault and explain my old student ties, to which the young woman replies, "Well, you should have plenty of fun, 'cuz they're already a couple of 'em holed up here."

Intrigued, I ask who.

"Bill and Harley're just down the block from you in sixteen and seventeen." Grinning, it dawns on me that I am not the only one to whom the topsoil at Hell Creek has become more like concrete. Only Bill's entourage of students is camped out in the breaks.

But it's not as though everything's changed, far from it. Fellman's pipes still reverberate with familiar rattlings that I became accustomed to as a student when we'd come in to shower and shop. More than a few facades on Main Street are almost reflexively recognizable, including John FitzGerald's drugstore, bathed in the blistering dry heat of the afternoon. Yet most reassuringly, the Hell Creek Bar still stands guard, emblazoned with its flaming logo, across the street on the corner. The rooms at Fellman's have been modestly redecorated, and the bar at the Hell Creek has been slightly repositioned. But John's sodas still tickle the tongue, and Joe's brews still numb a hot head.

Upon connecting with Harley and Bill, I sense in my viscera how much we've evolved. More wrinkles adorn all our faces and the pace of our outings has slowed. But our passion for this place has not ebbed, and we eagerly start catching up.

Despite finding himself a bit under the weather, Harley,

who last year turned eighty, seems to be dressed in the same blue work shirt, jeans, and cream-colored cowboy hat that he wore when I last "beat the bush" with him. With a somewhat fatigued yet resolute gaze, he plots out the next day's foray with Bill. The deep resonant tones projecting from behind Bill's full gray beard indicate that the ants in the area are in serious trouble. We'll return to the Engdahls to prospect for "mice." Of course I'm invited, but breakfast's at six.

In the "cool" of the morning, I head out to Engdahl's, following the same route I took on my first trip as a student. Though a few names have changed on the whitewashed slats on the sign at the critical crossroad, I don't need to stop, for I now know the way.

Out on the outcrop I find Bill and Harley harassing the anthills with studied intensity. Harley magnanimously offers up his "cheaters," but I decline, knowing they'll be much more productive for him. Within minutes the popping of a motorcycle bounces off the nearby buttes, and a lean, lanky figure rides across the ridge. The silhouette is all I need to see. It's Bob, out in search of his prodigal cattle. As he drops down the coulee to check out the trespassers, our greetings are acknowledged with grins.

"Hope I'm not too late for branding," I quip.

"If I'd a known we'd a saved a few," he fires back with a chortle.

Over the next few days it begins to sink in. I'm back home in the field at Hell Creek. But that home is not all it once was. Lester has passed away, and neither Cathy nor Duane still resides there. The Twitchells still roam their ranch off to the east, but many of the spreads have changed hands. No longer do the Hausos or MacDonalds hold sway next to the Engdahls, and some of the ranchers now charge for access. The $7.5-million sale of "Sue," a monstrous tyrannosaur found a decade ago in South Dakota, has changed some relationships between bone diggers and land owners. Most still provide generous access, but some require fees for collecting. To some extent, a free market for fossils has descended on Jordan.

During Harley's annual party at the Hell Creek Bar fifty of his friends wander in. He's not lost his touch with the locals, to which the pile of homemade pies and cakes clearly attests. I catch up with many old friends and learn of other environmental alterations. A few of the locals aren't very optimistic about Jordan's chances. The population continues to decline, as children like Cathy and Duane set sail for distant pastures. Many ranches are getting larger, with fewer people to run them. Swarms of Canadian combine crews annually descend like locusts on many of the fields to harvest what the locals once reaped themselves. Yet a fierce independence persists, and skepticism toward the government still looms.

The concept of the Big Open Great Plains Wildlife Range is no more popular today than when it was originally proposed. Nonetheless, with a tone of disgust entwined in his words, Bob informs me that vast acres of the prairie in the region have fallen under the jurisdiction of a government program called the Conservation Reserve Program or CRP. According to information published by the United States Department of

Agriculture and the Commodity Credit Corporation, which oversee the program, it is

> the Federal Government's single largest environmental improvement program . . .
>
> Established in 1985, the CRP encourages farmers to voluntarily plant permanent areas of grass and trees on land that needs protection from erosion, to act as windbreaks, or in places where vegetation can improve water quality or provide food and habitat for wildlife. The farmers must enter into contracts with the CCC lasting between 10 and 15 years. In return, they receive annual rental payments, incentive payments for certain activities, and cost-share assistance to establish the protective vegetation.
>
> Today, the CRP safeguards millions of acres of American topsoil from erosion: and by reducing water runoff and sedimentation it protects groundwater and helps improve countless lakes, rivers, ponds and streams.[1]

As a city boy living in New York since 1987, I have never even heard of this enormous federal program, but the ranchers of Garfield County certainly have. USDA records show that the program got off to a modest start in the Hell Creek region during 1992, when two contracts were signed covering less than three thousand five hundred acres.[2] The acreage devoted to the CRP in the county gradually rose through 1996. But in 1998 a whopping one hundred and twenty-two new contracts were signed, covering almost forty thousand acres after the initiation of the "New CRP" program. This modified USDA program stipulates, "Only the most environmentally sensitive land, in relation to its cost, is now accepted into the program, making optimum use of each taxpayer dollar to improve the environment." In 2003 a total of more than 83,500 acres of Garfield County ranch land is under the jurisdiction of the CRP, at rental fees averaging just under thirty dollars an acre. Still, that's just slightly less than 3 percent of the county's land. If one includes the CRP acreage in the adjacent counties of McCone and Musselshell to the east and west, the total balloons to more than a quarter of a million acres, or a little more than four hundred square miles. In essence, it seems that the government and the ranchers have come to a sort of financial compromise designed to return more of the prairie to a more natural state.

Both wildlife and livestock still dot the prairies and breaks. Although fewer ranchers run herds of sheep, cattle are commonly seen in the pastures, and even a few herds of domesticated bison wander the grasslands on a small number of ranches. As I drive the dirt roads and highways I encounter numerous herds of pronghorn, ranging in size from three to twenty-five. Occasional foxes, coyotes, deer, snakes, and raptors also still pop into view.

But while I wander through Hell Creek, the big news does not involve man-made alterations of the landscape. At dusk after one torrid day familiar thunderheads roll in from the northwest. Zeus throws a tantrum as Joe and I watch from the sanctuary of the bar. Such outbursts represent a mixed blessing. Inevitably, the first topic raised in any conversation with the locals is rain, or, more commonly, the lack of it. But Joe is instantly nervous. He's a former smoke-jumper and foresees flames in Hell Creek's immediate future. As I leave the bar I drive to a clearing to watch nature put on its show; bolts flash every three seconds, illuminating low buttes on the horizon.

I retire for the night without thinking much of it. I've seen such scenes many times before on past trips. One summer in the 1980s, lightning from a similar tempest ignited a small fire in the breaks north of Engdahl's. With Bob in the lead, we loaded up the shovels, rounded up the rancher to the north, and set out across the forested buttes to do what we could to control it. It was an intimidating and dangerous business, but my own role ended up resembling more a cameo with the Keystone Kops, rather than a starring role as a smoke-jumper. I was riding on the back bumper of the neighbor's pickup when, in his haste to get to the blaze, he drove over a protruding sapling, trapping the truck. As Bob hastened ahead and got down to work, we were stuck and couldn't get free. Eventually Bob pulled us out and we rushed to the front to help out. But by then the small blaze was just smoldering, so, with a wary eye on the fire we leaned back and counted satellites, "sputinks" as Bob preferred to call them, lazily crossing above.

Returning to 2003, I set out the next morning to photograph Black Butte, where Miles intended to corner Sitting Bull. A small plume of gray smoke lingers on the northern horizon. Joe had good reason to be fearful.

Over the next couple of days several small fires join forces, generating a massive blaze to the northwest of town. After starting in the breaks, they are driven by gusty winds across the prairies and ranchland to the south. An intimidating layer of dense, acrid smoke tints the town and surrounding landscape in unnatural tones of pale, hazy orange. Harley's party at the Hell Creek Bar is momentarily interrupted by the arrival of several out-of-state fire crews. With the entrance of a crew of Native Americans, an ethereal sense of unease wafts around the bar. Some locals have never forgotten the slaughter of Custer, and some also suspect that the government is intentionally letting the fire burn, ruining their neighbors' land in the name of ecological renewal. But Joe simply serves up the burgers and chicken, while talk of the fires continues.

Hell Creek is again in the news, but not for its frustrated Freemen. On July 23, across the bottom of the front page topped by the killing of Saddam's sons, the *Billings Gazette* reports what I'd heard the previous night. True to the region's long-standing spirit, neighbors responded to the imminent threat by banding together in brigades to help each other try to fend off the fires, which had now enveloped 117,570 acres. Accounts in the Hilltop Café, where we'd often eaten as students, varied as to how many buildings and how much livestock had perished. As I joined Harley and Bill for breakfast, our own table benefited from an informed assessment by Jerry Twitchell, who works on a ranch in the fire zone. Most of the damage is occurring in the breaks, but several structures are threatened. Critics of the government's Bureau of Land Management and other agencies involved argued that the fire could have been stopped before it raged out of control. The *Gazette* quoted one rancher in the area as saying, "It got this size because we were told not to use a cat [bulldozer] on about a twenty-acre fire, and it got away because of the federal bureaucracy." Old grudges step up to the fore.[3]

On July 24, as I take leave of Jordan to wander out west and visit the National Bison Range in Moiese, the highway is enveloped in a smoky fog. Over the next few days I keep up on events through the newspapers. The *Gazette* reports that the government has enlisted the help of two "Super Scoopers." These fire-fighting aircraft can "skim [across] the surface of Fort Peck Lake, scooping up 2,000 gallons of water in about 10 seconds" and deliver it to the fire lines.[4] The planes are on loan from Minnesota, where they are based. Other, larger tankers for ferrying water and fire retardant are based in Billings, Lewistown, and Miles City, but they can't make multiple runs to the fire like the "Super Scoopers."

The next day, as a fire near Glacier National Park begins to steal the headlines, the *Gazette* mixes the good news with the bad.[5] The fires have now burned 131,000 acres, but the firefighters, laboring in temperatures over one hundred degrees, estimate that the blaze is between 40 and 50 percent contained. Officials anticipate encircling the inferno within two or three days, which will allow them to begin transferring fire fighters to other more threatening blazes. Costs are approaching $1.75

million, with the final bill possibly reaching $3.5 million, of which the federal government would ante up 75 percent.

By July 27, after I returned to New York, officials report that the conflagration was 95 percent encircled by fire lines. The *Associated Press* elaborates that the flames claimed eight structures, including two old homesteads and several outbuildings. The cost of the efforts to fight the blazes stands at $2.4 million. Montana's governor travels to Jordan to meet with the citizens and fire crews.

> Ranchers wanted to know why the state does not own a CL-215 "Super Scooper" plane . . . Martz said the planes cost $28 million and it has been more cost-effective up to this point to hire the planes when needed.[6]

I doubt that the locals were mollified, despite the governor's praise for the community's efforts and her assurance that the state had done everything in its power to minimize the damage:

> "I understand that farming and ranching is your heritage, and this is important to you," she said. "We will do what we can to help."[7]

On the 28th the *Gazette* prints a brief summary of the fire's statistics, based on another *Associated Press* release, noting that the fire is 100 percent contained and that this would be the last report on Hell Creek's most recent act in the spotlights of the media's circus.[8]

Exactly what all will emerge from the smoldering coals of the pungent pines and prairie grass is obscure to my myopic eyes. Yet, two effects are acutely in focus. My personal ties to the land and the people, both present and past, have risen, phoenixlike, from the ashes, and I have no intention of letting them once again wither. But surpassing the scale of my singular soul, as the flames and frenzy abate, I can't help but reflect that Hell Creek's sleepy landscape and all its intrepid inhabitants are evolving in front of my eyes.

NOTES

1. Badlands and Branding.

1. For all references in this book to mythology:

   Hamilton, E. 1942, *Mythology: Timeless Tales of Gods and Heroes*. New York: Warner Books.

   Guerber, H. A. 1893. *Myths of Greece and Rome*. New York: American Book Co.

2. Time Traveling Through Hell Creek.

1. Alt, D., and D. W. Hyndman *Roadside Geology of Montana*. Missoula, Mont.: Mountain Press Publishing. 1986. This is the reference for discussions of plate tectonics and diatremes.

2. Ibid.

3. Early Exploration.

1. Ambrose, S. E. 1996. *Undaunted Courage: Meriwether Lewis, Thomas Jefferson, and the Discovery of the American West*. New York: Simon and Schuster: 56–57.

2. Ibid.

3. Ambrose, S. E. 1996: 94.

4. DeVoto, B. 1953. *The Journals of Lewis and Clark*. Boston, MA: Houghton Mifflin Co.: 483.

5. Jewett, T. O. 2000. A short study detailing Jefferson's influence on the then infant science of paleontology in the United States. earlyamerica.com: 1–4.

6. DeVoto, B. 1953: 105.

7. Ibid.

8. DeVoto, B. 1953: 106.

9. Garfield County Historical Society, J. McRae, president. 1999. *Trailin' Through Time*. Miles City, MT: Star Printing Company: 4–7.

10. Moulton, G. E., ed. 1983–1987. *The Journals of the Lewis and Clark Expedition*. Lincoln, NE: University of Nebraska Press: Volume 4: 130.

11. Moulton, G. E., ed. 1983–1987. Volume 4: 132.

12. Moulton, G. E., ed. 1983–1987. Volume 4: 141.

13. Moulton, G. E., ed. 1983–1987. Volume 4: 151.

14. Moulton, G. E., ed. 1983–1987. Volume 4: 152.

15. Moulton, G. E., ed. 1983–1987. Volume 4: 156–157.

16. Moulton, G. E., ed. 1983–1987. Volume 4: 159.

17. Moulton, G. E., ed. 1983–1987. Volume 4: 160.

18. DeVoto, B. 1953: 113.

19. DeVoto, B. 1953: 114.

20. Moulton, G. E., ed. 1983–1987. Volume 8: 226.

4. In the Wake of Custer's Ruin.

1. Ambrose, S. E. 1996: 154 (see chap. 3, n. 1).

2. Ambrose, S. E. 1996: 154.

3. Ibid.

4. Greene, J. A. 1991. *Yellowstone Command: Colonel Nelson A. Miles and the Great Sioux War 1876–1877*. Lincoln, NE: University of Nebraska Press.

5. Greene, J. A. 1991: 3–4.

6. Connell, E. S. 1984. *Son of the Morning Star*. San Francisco: North Point Press: 112–113.

7. Connell, E. S. 1984: 118–119.

8. Utley, R. M. 1994. *Little Bighorn Battlefield: A History and Guide to the Battle of the Little Bighorn*. Washington, DC: Division of Publications, National Park Service: 20.

9. Connell, E. S. 1984: 246.

10. Utley, R. M. 1994: 35.

11. Ibid.

12. Connell, E. S. 1984: 259.

13. Connell, E. S. 1984: 217.

14. Connell, E. S. 1984: 279.

15. Connell, E. S. 1984: 286–287.

16. Utley, R. M. 1994: 72–73.

17. Utley, R. M. 1994: 75.

18. Utley, R. M. 1994: 82.

19. Greene, J. A. 1991: 37.

20. Ibid.

21. Greene, J. A. 1991: 71.

22. Greene, J. A. 1991: 67.

23. Garfield County Historical Society, J. McRae, president. 1999: 7 (see chap. 3, n. 9). An alternative account of this battle, based on an oral history provided by an Assiniboin warrior named Strong Bear, is documented in *Cowboy Poetry, Classic Rhymes and Prose by D. J. O'Malley, The N Bar N Kid White*, 2000, edited by Janice and Mason Coggin, Dallywelter Press, Great Falls, MT: 175–178.

24. Greene, J. A. 1991: 83.

25. Greene, J. A. 1991: 88–89.

26. Greene, J. A. 1991: 93.

27. Greene, J. A. 1991: 94–96.

28. Greene, J. A. 1991: 107.

29. Greene, J. A. 1991: 121.

30. Greene, J. A. 1991: 131–132.

31. Greene, J. A. 1991: 134.

32. Greene, J. A. 1991: 138.

33. Greene, J. A. 1991: 154.

34. Utley, R. M. 1994: 82.

5. The Vanishing Herds.

1. Connell, E. S. 1984: 217 (see chap. 4, n. 6).

2. DeVoto, B. 1953: 98 (see chap. 3, n. 4).

3. DeVoto, B. 1953: 103.

4. DeVoto, B. 1953: 108.

5. DeVoto, B. 1953: 107.

6. DeVoto, B. 1953: 120–121.

7. Catlin, G. 1841. Letters and Notes on the Manners, Customs, and Condition of the North American Indians; Published by the author at the Egyptian Hall, London; Vol. I: 199–200.

8. Lepley, J. G. and S. Lepley, 1992. *The Vanishing West: Hornaday's Buffalo*. Fort Benton, MT: The River and Plains Society, 29–30.

9. Lepley, J. G. and S. Lepley. 1992: 71.

10. Lepley, J. G. and S. Lepley. 1992: 71–72.

11. Hornaday, W. T. 1925. *A Wild-Animal Round-Up*. New York: Charles Scribner's Sons: 5–6.

12. Hornaday, W. T. 1925: 6.

13. Hornaday, W. T. 1925: 7.

14. Hornaday, W. T. 1925: 9–10.

15. Hornaday, W. T. 1925: 12.

16. Hornaday, W. T. 1925: 13–14.

17. Connell, E. S. 1984: 137.

18. Hornaday, W. T. 1925: 18–19.

19. Hornaday, W. T. 1925: 19.

20. Hornaday, W. T. 1925: 25.

21. Hornaday, W. T. 1925: 29, 32.

22. Hornaday, W. T. 1925: 47–51.

23. Lepley, J. G. and S. Lepley. 1992: 42–43.

24. Lepley, J. G. and S. Lepley. 1992: 84.

25. Lepley, J. G. and S. Lepley. 1992: 86.

26. Lepley, J. G. and S. Lepley. 1992: 37.

27. Lepley, J. G. and S. Lepley. 1992: 86.

6. The Resurrection of *Tyrannosaurus*.

1. Hornaday, W. T. 1925: 55 (see chap. 5, n. 11).

2. Jordan, A. J. 2003. *Jordan*. Missoula, MT: Mountain Press Publishing Co.: 25. This book was first published in 1984.

3. Jordan, A. J. 2003: 78.

4. Jordan, A. J. 2003: 91.

5. Jordan, A. J. 2003: 95.

6. Jordan, A. J. 2003: 96.

7. Jordan, A. J. 2003: 97.

8. Jordan, A. J. 2003: 102.

9. Jordan, A. J. 2003: 115.

10. Jordan, A. J. 2003: 138.

11. Ibid.

12. Jordan, A. J. 2003: 139.

13. Hornaday, W. T. 1925: 79–80.

14. Hornaday, W. T. May 29, 1902. American Museum of Natural History, Department of Paleontology, Field Correspondence. New York, NY.

15. Brown, B. June 17, 1902. American Museum of Natural History, Department of Paleontology, Field Correspondence.

16. Ibid.

17. Ibid.

18. Brown, B. July 7, 1902. American Museum of Natural History, Department of Paleontology, Field Correspondence.

19. This extract and the previous two are from Jordan, A. J. 2003: 139–140.

20. Brown, B. July 12, 1902. American Museum of Natural History, Department of Paleontology, Field Correspondence.

21. Brown, B. August 12, 1902. American Museum of Natural History, Department of Paleontology, Field Correspondence.

22.   Ibid.

23.   Brown, B. September 3, 1902. American Museum of Natural History, Department of Paleontology, Field Correspondence.

24.   Brown, B. October 13, 1902. American Museum of Natural History, Department of Paleontology, Field Correspondence.

25.   Brown, B. June 5, 1905. American Museum of Natural History, Department of Paleontology, Field Correspondence.

26.   Brown, B. June 24, 1905. American Museum of Natural History, Department of Paleontology, Field Correspondence.

27.   Brown, B. July 15, 1905. American Museum of Natural History, Department of Paleontology, Field Correspondence.

28.   Brown, B. August 8, 1905. American Museum of Natural History, Department of Paleontology, Field Correspondence.

29.   Brown, B. August 22, 1905. American Museum of Natural History, Department of Paleontology, Field Correspondence.

30.   Brown, B. July 8, 1908. American Museum of Natural History, Department of Paleontology, Field Correspondence.

31.   Ibid.

32.   Brown, B. July 15, 1908. American Museum of Natural History, Department of Paleontology, Field Correspondence.

33.   Osborn, H. F. July 30, 1908. American Museum of Natural History, Department of Paleontology, Field Correspondence.

34.   Brown, B. August 1, 1908. American Museum of Natural History, Department of Paleontology, Field Correspondence.

35.   Osborn, H. F. August 10, 1908. American Museum of Natural History, Department of Paleontology, Field Correspondence.

36.   Brown, B. August 10, 1908. American Museum of Natural History, Department of Paleontology, Field Correspondence.

37.   Norell, M. A.; E. S. Gaffney and L. Dingus. 1995. *Discovering Dinosaurs in the American Museum of Natural History*. New York: Alfred A. Knopf, Inc.: 117.

38.   Norell, M. A., E. S. Gaffney and L. Dingus. 1995: 192.

39.   Norell, M. A., E. S. Gaffney and L. Dingus. 1995: 190.

7. Following in the Footsteps.

1. Horner, J. R. and D. Lessem. 1993. *The Complete T. rex.* New York: Simon and Schuster: 66.

2. Garbani, H. May 19, 2003. Correspondence to L. Dingus, New York, NY.

3. Horner, J. R. and D. Lessem. 1993: 67.

4. Garbani, H. 1966. Field Notebook. Harley Garbani and the Natural History Museum of Los Angeles County, CA.

5. Dingus, L. July 16, 2003. Notes on conversation with H. Garbani. L. Dingus, New York, NY.

6. Garbani, H. May 19, 2003. Correspondence to L. Dingus, New York, NY.

7. Clemens, W. A. April 7, 2003. Correspondence to L. Dingus, New York, NY.

8. Horner, J. R. and D. Lessem. 1993: 67.

8. Fatal Impact and Enormous Eruptions.

1. Alvarez, L. et al. 1980. Extraterrestrial cause for the Cretaceous–Tertiary extinction. *Science*, 208: 1095–1108.

2. Dingus, L. and T. Rowe. 1998. *The Mistaken Extinction: Dinosaur Evolution and the Origin of Birds.* New York: W. H. Freeman and Co: 332 pp. This book and references listed therein constitute the sources for the scientific information discussed in this chapter.

3. Sadler, P.M. 1981. Sediment accumulation rates and the completeness of stratigraphic sections. *Journal of Geology*, 89: 569–584.

4. Dingus, L. 1984. Effects of stratigraphic completeness on interpretation of extinction rates across the Cretaceous–Tertiary boundary. *Paleobiology*, 10: 420–438.

5. Alvarez, W. 1997. *T. rex and the Crater of Doom.* Princeton, NJ: Princeton University Press: 142–144.

9. Echoes of the Wild West.

1. Egan, T. April 23, 1996. Siege is subplot in town's survival drama. *New York Times*, New York, NY.

2. Ostrom, C. M. and B. A. Serrano. May 7, 1995. Land of the Freemen—"Republic of Montana." *Seattle Times*, Seattle WA.

3. Ibid.

4. Ibid.

5. Ibid.

6. Ibid.

7. Florio, G. March 18, 1996. Isolated town fears Freemen, confrontation. *Billings Gazette*, Billings, MT.

8. Bender, M. March 26, 1996. FBI moves on freemen. *Billings Gazette*, Billings, MT.

9. Florio, G. March 18, 1996.

10. Ibid.

11. Bender, M. March 27, 1996. Freemen's neighbors back U.S. *Billings Gazette*, Billings, MT.

12. Ibid.

13. Johnson, C. April 12, 1996. Journalists keep Jordan folks busy. *Billings Gazette*, Billings, MT.

14. McLaughlin, K. March 29, 1996. Reporters gather at Hell Creek Bar. *Billings Gazette*, Billings, MT.

15. McLaughlin, K. March 29, 1996. Reporters gather at Hell Creek Bar. *Billings Gazette*, Billings, MT.

16. cnn.com June 15, 1996. In Montana town, life settles back to normal. CNN: europe.cnn.com.

17. Florio, G. March 18, 1996.

18. Linsalta, P. April 21, 1996. Montana townspeople have had enough of the Freemen. *Detroit News*, Detroit, MI: detnews.com.

19. Bender M. April 4, 1996. State busy for networks. *Billings Gazette*, Billings, MT.

20. Nottingham, N. April 4, 1996. Will all the recent news darken the Big Sky? *Billings Gazette*, Billings, MT.

21. Crisp, D. April 8, 1996. Sense of humor alive in Jordan. *Billings Gazette*, Billings, MT.

22. Pankratz, H. May 31, 1996. FBI steps indicate move on Freemen nearing. *Denver Post*, Denver, CO.

23. Johnson, C. June 14, 1996. FBI convoy hauls fugitives away to Billings. *Billings Gazette*, Billings, MT.

24. Bender, M. March 27, 1996 (see n. 7).

25. Crisp, D. April 7, 1996. Balancing loyalties; Dispatcher's sister on Freemen farm. *Billings Gazette*, Billings, MT.

26. Billings, E. P. June 16, 1996. Clark plows land at last; reconciliation yet to be sown. *Billings Gazette*, Billings, MT.

27. Billings, E. P. June 14, 1996. Jordan gets back to normal. *Billings Gazette*, Billings, MT.

28. cnn.com. June 15, 1996. "In Montana town, . . ."

29. "Montana town still healing rifts five years after Freeman standoff," Associated Press, June 13, 2001.

30. Johnson, C. April 12, 1996 (see n. 8).

31. Egan, T. April 23, 1996 (see n. 1).

32. Ibid.

10. Is the Past the Key to the Future?

1. Mayne, J. July–August 1998. Nick Murnion: A profile in courage. *Rural Telecommunications*.

2. Egan, T. April 23, 1996 (see chap. 9, n. 1).

3. Ibid.

4. Ibid.

5. Baker, D. 1992. *Next Year Country*. Boulder, CO: Fred Pruett Books: 49.

6. Duncan, D. 1993. *Miles From Nowhere: Tales From America's Contemporary Frontier*. New York: Penguin Books: 31–32.

7. Turner, F. J. 1893. The significance of the frontier in American History, *Annual Report of the American Historical Association*: 199–227.

8. Duncan, D. 1993: 32.

9. Duncan, D. 1993: 33.

10. Alwin, J. A. 1981. Jordan country—A golden anniversary look. *Annals of the Association of American Geographers*, 71: 496.

11. Egan, T. May 27, 2001. As others abandon plains, Indians and bison come back. *New York Times*, New York.

12. Duncan, D. 1993: 56–57.

13. Duncan, D. 1993: 57–58.

14. Duncan, D. 1993: 59.

15. Ibid.

16. Sidey, H. September 24, 1990. Hugh Sidey's America: Where the buffalo roamed. *Time*.

Epilogue

1. Farm Service Agency, U.S. Department of Agriculture. 2003. The Conservation Reserve Program. http://www.fsa.usda.gov/in/ crp. htm.

2. Farm Service Agency, U.S. Department of Agriculture. July 31, 2000. Conservation Reserve Program Reports. http://www.fsa. usda.gov/crpstorpt/07Approved/r1landyr/mt.htm.

3. Bohrer, B. July 23, 2003. Hot winds threaten to revive fires. *Billings Gazette*, Billings, MT.

4. Gazette Staff. July 24, 2003 "Super Scoopers" now part of Breaks-area fire arsenal. *Billings Gazette*, Billings, MT.

5. Gazette Staff. July 25, 2003. Glacier fire forces evacuations. *Billings Gazette*, Billings, MT.

6. Associated Press (as printed in *Billings Gazette*). July 27, 2003. Plans laid for Glacier evacuations. *Billings Gazette*, Billings, MT.

7. Ibid.

8. Associated Press (as printed in *Billings Gazette*), July 28, 2003. Weather slows fire's move toward park offices. *Billings Gazette*, Billings, MT.